CLASSIFICATION OF HAZARDOUS LOCATIONS

D1806609

By A.W. Cox, F.P. Lees and M.L. Ang

A report of the Inter-Institutional Group on the Classification of Hazardous Locations (IIGCHL)

Member Institutions:
The Institution of Chemical Engineers
The Institution of Electrical Engineers
The Institution of Gas Engineers
The Institution of Mechanical Engineers

INSTITUTION OF CHEMICAL ENGINEERS

i

Published by
Institution of Chemical Engineers,
Davis Building,
165–189 Railway Terrace,
Rugby, Warwickshire CV21 3HQ, UK.
IChemE is a Registered Charity

© 1990 Institution of Chemical Engineers
Reprinted May 1991
Reprinted February 1993
Reprinted May 1996
Reprinted April 1998
Reprinted January 2000
Reprinted May 2003
ISBN 0 85295 258 9

Printed and bound by Antony Rowe Ltd, Eastbourne

EXECUTIVE SUMMARY

The work described in this report is essentially an investigation of the feasibility of putting hazardous area classification on a more quantitative basis. The study was sponsored by the Inter-Institutional Group on the Classification of Hazardous Locations (IIGCHL) which was set up to promote an inter-disciplinary approach to area classification.

Current guidance is inconsistent, and leads to problems in defining zones particularly on indoor plant. There are considerable differences in some of the zone distances recommended in the various codes; examples of these are given. Figure 3.1 on page 14 illustrates the differences in zone distances for LPG.

The main thrust of the work has been to develop concepts and methods which could form the basis of a more quantitative approach to area classification, but no attempt has been made to produce an actual design model.

The report includes reviews of the development of the subject, current standards and codes and of current performance in the control of ignition sources, and a description of the risk criteria which may be used in a quantitative treatment. Risks estimated are expressed primarily in terms of the Fatal Accident Rate (FAR).

The risk to workers from ignition of flammable leaks was reviewed. Although the data on which to make a judgement are sparse, the risk is higher than originally anticipated. The risk estimated for plant workers in the oil and chemical industries is an FAR of 0.56% per 10^8 working hours.

Basic information and models relating to a quantitative approach were collected. These included inventories of leak sources on plants and the national inventory of plants; a set of standard hole sizes similar to those currently used in industry; estimates of leak frequency and hole size distribution obtained from equipment failure data; a set of models for emission, vaporisation and dispersion of leaks and properties of a set of representative flammable fluids; and estimates of the probabilities of ignition and explosion.

The problem was approached in two stages. The first stage was the development of a method to compare the hazard ranges or zone distances given in different codes, on the basis of the different factors which affect these ranges (ie the fluid properties and operating conditions) using the standard hole sizes, emission and dispersion models and the representative flammable fluids.

The second stage involved the development of a method of applying the risk assessment techniques to obtain an estimate of the absolute risk implied by a particular zoning policy. This more ambitious aim proved difficult to achieve because information on leaks and ignitions is hard to obtain. What has been done is to devise an overall approach and to go part way to implementing it.

In order to make a risk assessment it is necessary to have estimates of the inventory of leak sources on a plant, the leak frequencies and hole size distributions and the ignition and explosion probabilities. Leak sources on plants were studied and a leak source profile for a 'standard' plant was derived. A fire and explosion model was developed in which the input data were the leak frequency and hole size distribution and the ignition and explosion probabilities, and the outputs were the frequency of leaks, fires and vapour cloud explosions, both overall and by fluid phase and by leak sources, such as pipework, valves, etc. Outputs from the models can be compared with the corresponding historical data. The purpose of the model is to assist in obtaining improved estimates of the leak frequencies and hole size distributions, and the ignition and explosion probabilities.

Attention is drawn to two points which are highlighted in the report. One is the importance of self-ignition of leaks, which may account for a relatively large proportion of small leak ignitions. Another is the importance of flashing liquids such as LPG. Such a fluid combines the high emission rate characteristics of a liquid with the dispersion features of a vapour and can give a long hazard range.

The work principally applies to outdoor plants, but the problems of hazardous area classification of indoor plants were also considered. Two important features of leaks in indoor plants are leaks other than those from fixed, closed plant, and the various aspects of ventilation. It is proposed that a method be developed for the identification of leak situations in indoor plant, such as those arising from operational or maintenance activities and from transfer of materials; a technique akin to hazard and operability (hazop) studies is envisioned.

An attempt has been made to define the situations in which a limited, as opposed to a blanket, zoning might be used for indoor plant. A precondition of limited zoning is that the situation be relatively predictable. Even then limited zoning appears applicable only in a rather restricted set of cases.

The question of minimum inventory was also considered, and the report concludes that a minimum inventory might be permitted under certain conditions.

CONCLUSIONS

The work described in this report is essentially an investigation of the feasibility of putting hazardous area classification on a more quantitative basis. The work did not reach the point where clear and uncontentious guidance on hazardous area classification can be given, but an overall type of approach to classification has been produced which includes models and risk criteria and which might form the basis of further work aimed at producing guidelines.

The authors believe this reporting of interim results, as opposed to a finalised method of hazardous area classification, to be useful, since it allows for comment from the profession before any further development of methodology.

FUTURE WORK

The study has identified the need for future work which might proceed broadly along the following lines: the effect of zoning policies would be investigated using a computer-based plant layout package. The inputs would be the layout of the standard plant, the leak frequencies and hole size distributions and the hazard ranges given by the emission and dispersion models. The zones defined by existing codes would also be an input. Estimates would then be made of the distribution of ignition sources both outside the zones and also inside them, the latter being due to failure completely to exclude such sources. These estimates would then be adjusted to give the historical ignition and explosion probabilities. Having obtained a model of the situation using present zoning policies, the effect of alternative policies could be explored. These policies could include definition and range of both Zone 1 and Zone 2.

It is expected that the risk assessment approach just described would be used for the production of general guidelines along the lines of existing codes, rather than for major extension of the source of hazard approach. With the latter appreciably more effort is required and uncertainty in the models and data used bulks larger.

The following recommendations have been made for future work:
1) Comparison of the different risk criteria available, so that the most suitable criterion can be selected and where necessary refined into a form suitable for hazardous area classification.
2) Refinement of the fire and explosion model.
3) Estimation of the probability of injury due to ignition of leaks.
4) Application of the refined fire and explosion model to zoning of outdoor plants.
5) Study of self-ignition of leaks.
6) Development of the concept of limited relaxations for outdoor plants, including criteria for deciding where such relaxations are appropriate and to what extent they should be applied.
7) Extension of the work to indoor plants. This would include:
a) development of hazard identification techniques for activities (as defined above) in indoor plant;
b) examination of the mechanisms of dispersion by ventilation;
c) identification of situations where quantification of zoning is applicable to indoor plants.
8) Application of quantified hazardous area classification concepts to offshore platforms.
It is suggested that items 1, 5, 7 (a) and (b) and 8 might be separate studies.

ACKNOWLEDGEMENTS

The authors would like to thank the organisations and companies which supported this work through the Inter-Institutional Group and to acknowledge the guidance and encouragement received from the Guiding Committee as a whole and from its individual members.

LIST OF SPONSORS
Associated Octel Company Ltd
Association of Consulting Engineers
British Gas Plc
British Petroleum Company Plc
Engineering Equipment and Materials Users Association (EEMUA)
Enterprise Oil Plc
Foster Wheeler Energy Ltd
Glaxochem Ltd
Health and Safety Executive
Humphreys & Glasgow Ltd
Institution of Chemical Engineers
Institution of Electrical Engineers
Institution of Gas Engineers
Institution of Mechanical Engineers
Insurance Technical Bureau
Matthew Hall Engineering Ltd
Snamprogetti Ltd

CONTENTS

1. INTRODUCTION

This report gives an account of a project sponsored by the Inter-institutional Group on the Classification of Hazardous Locations (IIGCHL) at the Department of Chemical Engineering, Loughborough University of Technology.

The aim of the project was the review of the overall approach to the classification of hazardous locations with particular reference to the possible application of a more quantitative approach. The terms of reference of the project are given in Appendix A.

An account of the background to the problem and of previous work in this area is given in Section 2. Current standards and codes are reviewed in Section 3 and current performance in control of ignition sources in Section 4. Section 5 deals with the objectives and criteria of the project and Section 6 with its strategy. Section 7 reviews leak sources and scenarios. In Sections 8–17 the quantitative data relevant to the problem are assembled. Section 8 gives an inventory of leak sources on a typical plant and Section 9 an estimate of the number of plants. Section 10 defines a set of standard hole sizes similar to those currently used in industry. Section 11 gives estimates of leak frequency and hole size distribution obtained from equipment failure data. A set of models for emission and dispersion and a set of representative flammable fluids are given in Sections 12 and 13, respectively. Section 14 describes a numerical investigation of the typical hazard ranges for these fluids. Sections 15 and 16 present estimates of the probabilities of ignition and of explosion, respectively. Section 17 reviews data on fires and on vapour cloud explosions. Section 18 gives a fire and explosion model, the purpose of which is to crosscheck and refine the estimates of leak frequency and hole size distribution and of ignition and explosion probability. Indoor plants are considered in Sections 19–21, which deal respectively with ventilation, general considerations and zoning. The overall approach to hazardous

1

area classification is given in Section 22 and conclusions and recommendations in Section 23.

The project took three years, from February 1986 to April 1989. It was overseen by the IIGCHL chaired by Mr G.A. Lee and by a steering committee chaired initially by Dr K. Ramsay and then by Mr C.A.W. Townsend. The composition of these two groups is given in Appendix B.

The work was carried out by Mr A.W. Cox under the guidance of Professor F.P. Lees and Dr M.L. Ang.

2. CLASSIFICATION OF HAZARDOUS LOCATIONS: BACKGROUND

2.1 DEVELOPMENT OF CODES

The development of the classification of hazardous locations has been described by Palles-Clark (1982). This development is summarised here, drawing on this account and bringing it up to date.

The need to control ignition sources was recognised early in the coal mining industry, where the presence of methane can lead to flammable atmospheres, and led to the development of flameproof equipment. In Britain the British Standards Institution published BS 229 Flameproof Enclosure of Electrical Apparatus in 1926.

The application of this concept to gases other than methane led in 1946 to a further edition of BS 229 which dealt with a range of gases and vapours.

Also in 1946 the Institute of Petroleum Code gave the first guidance to electrical engineers on hazardous areas, distinguishing between a 'dangerous area' and a 'safe area'.

The first BS Code of Practice covering the need to use flameproof and intrinsically safe equipment in industries other than coal mining appeared in 1948. Again it was based on the concept of a 'dangerous area' but did not define it except that it referred to the persistence of a flammable atmosphere.

In parallel with these developments in Britain other countries were also developing their own approaches to control of ignition sources. Much early work in this area was done by the American Petroleum Institute (API).

For the best part of 20 years electrical engineers had to make do with this concept of a dangerous area as the basis for the installation of flameproof or intrinsically safe equipment.

In 1965 the third edition of the IP Electrical Code referred to a 'remotely dangerous area', introducing the concept that a flammable atmosphere in such an area would be abnormal. For such areas the specification of the electrical equipment was relaxed to allow the use of non-sparking or low VA equipment. There was also reference to the limitation of surface temperature.

Between 1964 and 1967 BS CP 1003 was revised and reissued in three parts. The code referred to three hazardous areas, defined essentially in terms of the probability of the presence of a flammable atmosphere.

The working party included representatives from IChemE and this activity marks the start of a greater interest and contribution by chemical engineers in hazardous area work.

The early 1970's saw the issue of several new British Standards relating to hazardous areas:

BS 4683 1971 Electrical apparatus for explosive atmospheres
BS 5345 1976 Pt 1 Code of practice for the selection, installation and maintenance of electrical apparatus for use in potentially explosive atmospheres
BS 5501 1977 Electrical apparatus for potentially explosive atmospheres
It is BS 5345 Pt 2 Classification of hazardous areas which is of main relevance here.

In 1972 ICI issued its Electrical Installations in Flammable Atmospheres Code (the ICI Electrical Installations Code) (ICI/RoSPA 1972).

Over the years there has been considerable criticism of the approaches followed in these codes. In particular, the IEE has held a number of meetings on the subject, notably the four symposia on Electrical Safety in Hazardous Environments (IEE 1971, 1975, 1982a, 1988) and the colloquium on Flammable Atmospheres and Electrical Area Classification (IEE 1982b). A revision of the IP Electrical Safety Code is currently being undertaken, including revision of the provisions for hazardous area classification.

2.2 LEGAL REQUIREMENTS

In most work situations the Health and Safety at Work, etc, Act 1974 will apply. This requires, amongst other things, the provision and maintenance of plant and systems of work that are safe and without risks to health, so far as is reasonably practicable. This is a very general requirement which will apply to all aspects of avoidance of ignition of flammable gases and vapours, including ignition due to electrical equipment.

The main statutory requirements governing electrical installation and equipment are contained in the Electricity at Work Regulations 1989, which came into force on 1 April 1990. These Regulations replace the old Electricity (Factories Act) Special Regulations of 1908 and 1944, and apply to all places of work, and not just factories.

Regulation 6, on adverse or hazardous environments, states that: "Electrical equipment which may reasonably foreseeably be exposed to— (a) mechanical damage; (b) the effects of weather, natural hazards, temperature or pressure; (c) the effects of wet, dirty, dusty or corrosive conditions; or (d) any flammable or explosive substance, including dusts, vapours or gases, shall be of such construction or as necessary protected as to prevent, so far as is reasonably practicable, danger arising from such exposure."

In addition, the Highly Flammable Liquids and Liquefied Petroleum Gases Regulations 1972 state in Regulation 9.1: "No means likely to ignite vapour from highly flammable liquids shall be present when a dangerous concentration of vapours from highly flammable liquids may be reasonably expected to be present." In general, these Regulations will apply to premises where the Factories Act applies. A highly flammable liquid is one which has a flash point below 32°C and supports combustion under specified test conditions.

The Petroleum (Consolidation) Act 1928 requires the keeping of petroleum spirit and certain other materials to be licensed. The licence may have conditions attached to it which may relate to area classification matters.

2.3 DEFINITION OF HAZARDOUS AREAS

The international definitions for the zones used in hazardous area classification are as follows:

Zone 0   A zone in which a flammable atmosphere is continuously present or present for long periods.

Zone 1   A zone in which a flammable atmosphere is likely to occur in normal operation

Zone 2   A zone in which a flammable atmosphere is not likely to occur in normal operation and if it occurs will only exist for a short time.

A non-hazardous area is an area not classified as Zone 0, 1 or 2.

2.4   DISCONTENT WITH CURRENT APPROACHES

It is generally recognised that the present approaches to hazardous area classification are not based on any very scientific foundation. They are essentially empirical and possibly conservative.

It is probably fair to say that despite this most engineers involved do not see classification of hazardous areas as a serious problem. The present approach provides a practical system to which they can design. A degree of conservatism is considered appropriate.

Other engineers, however, feel that it should be possible to do better. Their criticisms tend to come under five heads.

First, the current approaches are excessively conservative and in some cases lead to unneccessary costs.

Second, they are sometimes inconsistent and occasionally lead to nonsense. Thus there are inconsistencies between the different codes. And there are nonsenses such as the use of safeguarded equipment despite the close proximity of an open flame.

Third, they are not adequate in respect of indoor plants.

Fourth, there is the feeling that it should be possible to make more effective use of quantitative methods.

Fifth, there is a special problem for the HSE in that in this situation it is not easy to assess proposals which are on the borderline of accepted practice.

From discussions held with the IIGCHL these appear to be the principal points of discontent with current approaches. The points are considered further below.

2.5   VIEWS ON QUANTIFICATION

There are a number of pieces of evidence that support exists for a more quantitative approach. These include the use of the source of hazard (SoH) method in a number of companies and codes, the work done on the quantitative appendix for BS 5345 and the present project. There is also, of course, opposition to this approach.

The main development in quantification in recent years has been the SoH method. In this method, instead of setting the zones using a blanket approach, zoning is based on estimated travel distances of leaks from potential leak sources to potential ignition sources. Three levels of source of hazard are defined: Source of hazard 0 (continuous), source of hazard 1 (primary) and source of hazard 2 (secondary). A detailed account is given in the ICI Electrical Installations Code (ICI/RoSPA, 1972).

In order to use the SoH method it is necessary to have the hole size for the leak and to be able to model the emission and dispersion. Generally fixed hole sizes are used for each particular type of equipment.

There are two views on the SoH method. One is that it is good, because it forces the engineer to look more closely at individual leak sources and thus to identify the significant sources. A closely related view is that it gives greater reassurance that the zone distances are right.

The SoH method does take account of the fluid and of the operating conditions which are, or should be, important features in zoning. It also permits allowance to be made for engineering features intended to reduce the risk of ignition.

The contrary view is that the SoH method has two major drawbacks. One is that it is much effort for little return in the sense that the zone distances, and overall zoning, finally selected tend to be much the same as they would be using the blanket method.

The other is that the uncertainties in respect both of leak frequency and hole size distribution and of dispersion modelling are such that little confidence can be placed in the prediction of travel distance for a specific leak situation.

In discussions of the desirability or otherwise of taking quantification further, quantification is often not well defined, but the quantitative approach which most people seem to have in mind is one based on classic risk assessment which would involve determining the leak frequency and hole size distribution, modelling the emission, vaporisation, dispersion and ignition of the leak and assessing the results against some absolute risk criterion.

Assuming that progress can be made in this direction, the results may be applied on two different ways. One is to refine the guidelines which underlie the blanket approach. The other is to refine the SoH method.

Another use of a quantitative approach might be to make it easier to handle special cases, where a literal interpretation of the blanket or SoH methods appears to give a result which engineering judgment suggests is overkill, but which at present has to be implemented.

The decisions which might be assisted by a quantitative method are not limited to those on zone distances. They may also include the decision on the class of zone. For ventilated systems there is the decision on the required level of ventilation and on ventilation system reliability.

2.6 TERMINOLOGY

In the following use is made of certain terms which it is convenient to define here.

A distinction is made between plants in the open air and plants in buildings. These are referred to as outdoor plants and indoor plants, respectively.

Another distinction made, which is applicable mainly to indoor plants, is between regular process plants in which the fluids are usually fully enclosed, and those in which open surfaces are a feature, such as spray booths, degreasing plants, etc. These are referred to as closed plants, or regular process plants, and plants with open surfaces, respectively.

For individual items of equipment leak sizes are referred to as rupture, major and minor. For plants leak sizes are classed as massive, major and minor. The terms major and minor have different meaning in respect of equipment and of plant. Thus a leak which is major for a small bore connection might be minor for a plant.

The term leak rate is avoided. Reference is made to leak frequency or leak flow, as the case may be. The term leak size is used in the sense of leak flow, the size of the leak aperture being termed hole size.

In presenting numerical values, it is convenient in some cases, for clarity and to avoid accumulation of rounding errors, to give results to several significant figures; there is no implication that the results are necessarily of that accuracy.

3. CLASSIFICATION OF HAZARDOUS LOCATIONS: CURRENT STANDARDS, CODES AND GUIDES

The development of standards and codes for classification of hazardous locations was outlined in Section 2, where mention was made of BS 229, BS CP 1003, BS 5345 and other British Standards, of the IP Code and of the ICI Code. In this section a summary is given some of the standards and codes in the UK and overseas.

A listing of some of the relevant standards and codes on hazardous area classification, for the UK and overseas, is given in Appendix 1.

Table 3.1 gives a list of UK standards, codes and guidance on hazardous area classification. Only one of the items, the British Gas standard, adopts an approach which may be regarded as fully quantitative. Perhaps a principal reason for this is that the guide is concerned only with methane, which results in considerable simplification of the range of systems to be considered, the calculational models, etc.

An attempt was made to provide BS 5345 Part 2 with quantitative methods of setting zone distances, but the way in which the guidance was presented and the method itself were the subject of criticism and and the appendix on quantitative methods was discarded.

Table 3.2 lists a selection of the guides available overseas.

Offshore operations are the subject of separate guides, some of which are listed in Table 3.3.

Some of the potential emission sources covered by the guides in Tables 3.1–3.3 are shown in Table 3.4. This table does not attempt to cover all the sources in each guide, but rather is intended to provide a broad indication of the coverage of the guides.

Table 3.1 Guidance on hazardous area classification: some UK guides

Organisation	Publication	Approach
British Standards Institution	BS 5345 Pt 2, 1983	Quantitative approach suggested, but no guidance given
Institute of Petroleum	Model Code of Practice, Part 1, 1965	Suggested fixed distances
ICI/RoSPA	Electrical Installations in Flammable Atmospheres, 1975	Suggested distances which vary with material
British Gas	Hazardous Area Classification for Natural Gas, BGC/PS/SHA1, 1986	Quantitative method
BP Chemicals	Area Classification, 1986	Suggested fixed distances, supported by quantitative method
Health and Safety Executive	Hazardous Area Classification: A Guide for Fire and Explosion Inspectors, 1986	Based on ICI/ RoSPA code supported by quantitative method
Health and Safety Executive	The Storage of Highly Flammable Liquids, CS 2, 1978	Suggested fixed distances
Health and Safety Executive	The Storage of LPG at Fixed Installations, CS 5, 1981	Suggested fixed distances
Health and Safety Executive	The Keeping of LPG in Cylinders or Similar Containers, CS 4, 1981	Suggested fixed distances
Health and Safety Executive	Highly Flammable Liquids in the Paint Industry, 1978	Suggested fixed distances
Health and Safety Executive	Vehicle Finishing Units: Fire and Explosion Hazards, PM 25, 1981	Suggested fixed distances
Health and Safety Executive	Bulk Storage and Handling of High Strength Potable Alcohol, 1986	Suggested fixed distances

Table 3.2   Guidance on hazardous area classification: some overseas guides

Organisation	Publication	Approach
American Petroleum Institute (US)	Classification of locations for electrical installations in petroleum refineries, API 500A, 1982	Suggested fixed distances
National Fire Protection Association (US)	Flammable and combustible liquids code, NFPA 30	Suggested fixed distances
National Fire Protection Association (US)	National Electrical Code, NFPA 70	Suggested fixed distances
Directorate-General of Labour (Netherlands)	Guidelines for the classification of hazardous areas in zones in relation to gas explosion hazards and to the operation and selection of electrical apparatus R 2, 1979	Suggested fixed distances based on given leak rates
Berufsgenossenschaft der chemischen Industrie (FRG)	Guidelines for the prevention of danger in explosive atmospheres with examples, 1975	Suggested fixed distances
Comité Professionel du Pétrole (France)	Rules for the management and usage of LPG storages, 1972	Suggested fixed distances
Comité Professionel du Pétrole (France)	Rules for the management and usage of liquid hydrocarbons, 1975	Suggested fixed distances
Comitato Elettrotecnico Italiano (Italy)	Code for electrical installations in locations with fire and explosion hazards, 64–2, 1973	Suggested fixed distances
Sweden	Classification of hazardous areas, SS 421 08 20, 1984	Suggested fixed distances
Australian Standards Association	Classification of hazardous areas, AS 2430, 1981	Suggested distances based on flammability

11

Table 3.3 Guidance on hazardous area classification: some guides for offshore installations

Organisation	Publication	Approach
Institute of Petroleum (UK)	Model Code of Safe Practice, Part 8, 1964	Suggested fixed distances based on Part 1
American Petroleum Institute (US)	Classification of areas for electrical installations at drilling rigs and production facilities on land and on marine fixed and mobile platforms, API 500B, 1973	Suggested fixed distances based on API 500A
International Maritime Organisation	Code for the construction and equipment of mobile offshore drilling units, MODU Code, 1980	Suggested fixed distances
Det Norske Veritas	Offshore installations technical note B302: area classification and ventilation, 1981	Suggested fixed distances based on API 500B

Table 3.4   Guidance on hazardous area classification: coverage of selected guides

Item	1	2	3	4	5	6	7	8	9	10	11	12	13	14
Heavier than air source	x				x	x	x	x	x	x	x	x		
Lighter than air source	x				x	x		x			x	x	x	
As above – enclosed and ventilated	x	x			x	x	x	x	x	x	x	x		
Storage tank – fixed roof and floating roof	x		x	x	x	x	x	x	x	x	x		x	
Relief valve/vent	x		x	x	x	x			x			x	x	
Sample point/drain			x						x					
Flanges/connections			x	x				x			x			
Tanker loading			x	x			x			x			x	
Separators			x		x		x			x				
Agitator gland			x											
Drum or small container filling			x				x	x	x				x	
Pumps/compressors			x	x		x	x		x	x		x	x	
Aircraft refuelling							x	x						
Paint spraybooth							x	x						
Petrol store	x													
Open liquid surfaces			x				x					x	x	
LPG store			x						x			x		
Garage/service station	x		x				x	x			x			

Drilling operations

Item	1	2	3	4	5	6	7	8	9	10	11	12	13	14
Well head equipment		x				x								x
Base of drill	x					x								x
Mud tank	x					x						x		x
Shale shaker						x						x		x
Christmas tree						x								x
Pumped well						x								x
Pig launching/collecting point						x						x		

Key:

1	IP Code Part 1	8	R No 2 (Netherlands)
2	IP Code Part 8	9	BCI Guidance Notes (FRG)
3	ICI/RoSPA Code	10	CCP Guidance Notes (France)
4	HSE Guidance Notes	11	CEI 64–2 (Italy)
5	API 500A (US)	12	TN B302 (Norway)
6	API 500B (US)	13	SS 421 08 20 (Sweden)
7	NFPA 30 and 70 (US)	14	MODU Code (IMO)

The recommendations of these guides vary appreciably. For example, Figure 3.1 shows the zoning recommendations for an LPG pump. Some guides do not differentiate between LPG and other flammable substances when setting zone distances and these are shown in the figure without designation. The difference between the smallest and largest zones is appreciable.

Figure 3.2 shows the zoning recommendations for a flange on LPG pipework.

Figure 3.3 shows the zoning recommendations for fixed roof storage tanks.

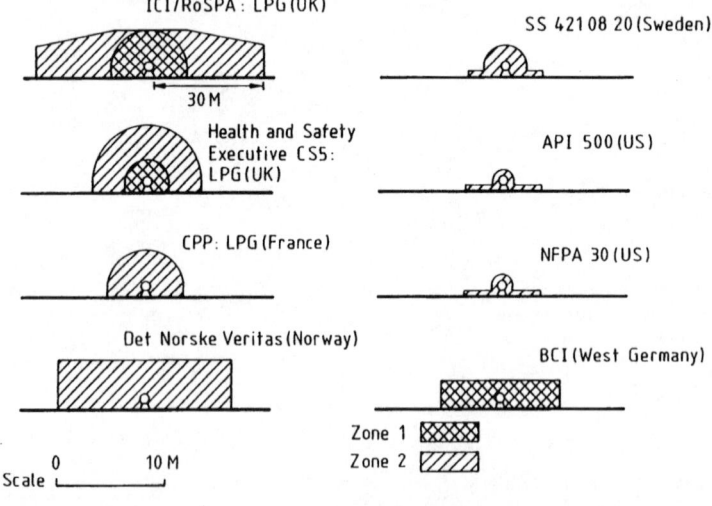

Figure 3.1   Zone 2 distances for pumps handling LPG given by selected standards and codes

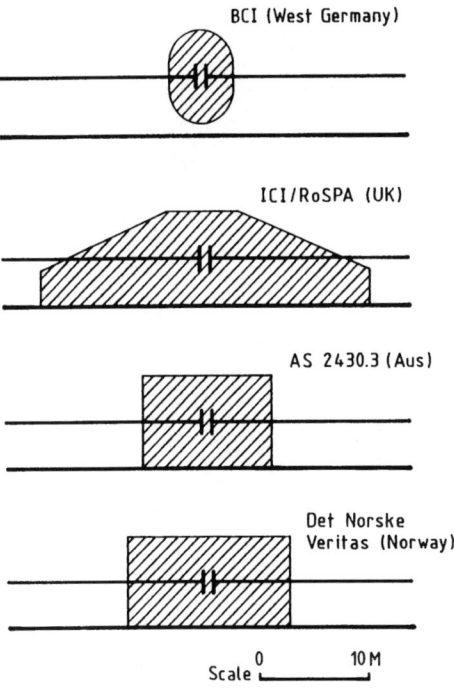

Figure 3.2   Zone 2 distances for flanges on pipework handling LPG given by selected standards and codes

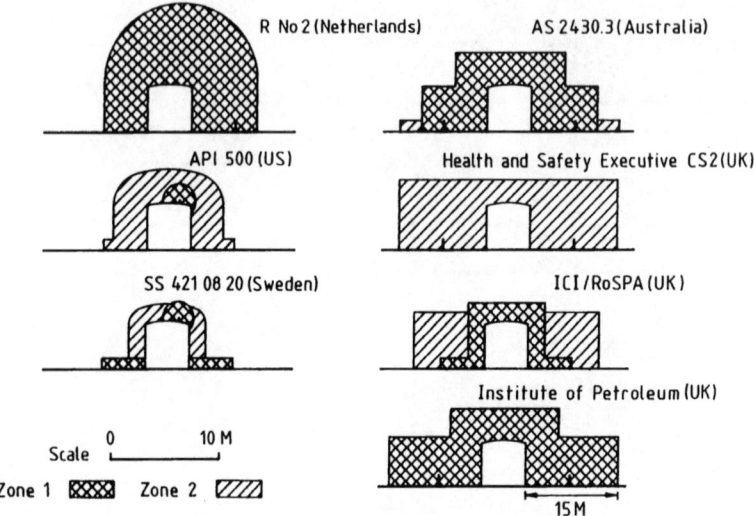

Figure 3.3   Zone distances for fixed roof storage tanks given by selected standards and codes

4. CLASSIFICATION OF HAZARDOUS LOCATIONS: CURRENT PERFORMANCE

4.1 FATALITIES AND INJURIES

There are statutory requirements for the reporting of accidents to the HSE. Information on these accidents has been made available to the project.

Arrangements for reporting of incidents to the HSE changed in March 1986 with the introduction of a new set of regulations, the Reporting of Injuries, Diseases and Dangerous Occurrences Regulations (RIDDOR). The incidents for which reporting is required include, but are not limited to, fires and explosions. Statistical information based on the new reporting arrangements was made available for the period April 1986 to March 1987. This fire and explosion survey is described in Appendix 2.

Separate treatment is given in this appendix to regular, essentially closed, process plant and to plant with open surfaces, etc.

An analysis of the data on fatalities and injuries for process plant in this survey is given in Appendix 3.

One important measure is the Fatal Accident Rate (FAR), which is the number of fatalities per 10^8 exposed hours. Safety performance may be measured by comparing the actual FAR with some target value. A target value which has been widely used is 0.4. An account of the Fatal Accident Rate and of target values is given in Section 5.

The estimate given in Appendix 3 for the actual FAR in the chemical and oil industries is 0.56.

It was anticipated at the start of the project that the risk of injury or fatality due to ignition incidents on regular process plant would be found to be so low as to be negligible. It appears from the estimates made that this may not be the case.

Although they are more difficult to estimate, the risks on indoor plants, both closed plants and, particularly, plants with open surfaces, etc, appear to be higher than those for outdoor process plants.

Some case histories of ignition of a flammable leak or atmosphere and of injury from such ignitions are given in Appendix 4.

It should be mentioned that the Abbeystead incident is not regarded as relevant to the present study and the 16 fatalities which occurred there have not been included in the statistics considered.

4.2   PROPERTY DAMAGE

As just mentioned, it was originally anticipated that risk of fatality or injury due to ignition incidents on outdoor plants would be found to be negligible and that attention would focus on property damage. Since this does not appear to be the case, the question of property damage has not been pursued in detail.

Information is available on fire losses. For example, the API issues annual fire loss reports for refinery fires in the US. For 1985 the breakdown of fires was as follows (API, 1985):

Size of loss (1000$)	No. of fires
2.5–100	65
100–1000	37
>1000	7

There have also been a number of surveys by insurers (eg Doyle, 1969; Spiegelman, 1969, 1980) which give the breakdown of property damage by fire and explosion. For large losses incidents, surveys tend to show that about one third of the losses are due to fire and two-thirds to explosions. Table 4.1 shows a breakdown given by Kletz (1977) of some data given by Doyle.

Table 4.1   Analysis of large loss fires and explosions (after Kletz, 1977)

	Proportion (%)
Explosions inside equipment because air got in	11
Explosions inside equipment because of runaway reaction or explosive decomposition	23
Explosions outside equipment but inside buildings	24
Explosions in the open	3
Vessels bursting (due to corrosion, overheating or overpressure)	7
Fires	32
Total	100

Information of this sort may be used to estimate the cost of leak ignition incidents. In the breakdown given by Kletz the explosions can be divided into some 27% which clearly involve ignition of a leak and some 41% which do not. The fires are less easy to classify. However, it seems a reasonable estimate that overall some 35% of large loss fires and explosions are leak ignition incidents.

5. CLASSIFICATION OF HAZARDOUS LOCATIONS: OBJECTIVES AND CRITERIA

5.1 OBJECTIVES

The overall objective of the work is to try to put the classification of hazardous locations on a more quantitative basis, with reference both to outdoor and indoor plants.

A quantitative approach based on classic risk assessment methods would proceed broadly as follows. The various leak sources are identified, the inventory of leak sources on a typical plant is determined, the leak frequency and leak size distribution of each source is estimated, the leak behaviour of a range of representative fluids is modelled, the probabilities of ignition and of explosion given ignition are estimated and the frequencies of fires and explosions per plant are calculated. Then from estimates of the probability of, say, fatality given fire or explosion, the frequency of such fatalities may be calculated. Likewise, from estimates of the degree of property damage from such fire and explosion, calculations may be done of the average property damage. Estimates are then made of the number of plants at risk and of the number of people employed and the national individual risk calculated. The national property damage is also calculated

The estimates so made may be crosschecked in various ways. For example, the frequency of fires and explosions should agree with historical data. But in addition, the proportion of fires and explosions attributable to leaks from particular types of equipment should also tie up. And so on.

This is the general approach adopted in risk assessment. There are, however, numerous treatments which fall short of this but which may nevertheless constitute a worthwhile quantitative approach to the problem and are more realistic from the viewpoint of the designer. They include the following:

1. Determination of the individual risk
2. Determination of the ignition source distribution
3. Determination of hazard ranges for different flammable fluids
4. Determination of the probability of ignition and explosion for different leaks.

The approach to quantification actually taken in this project is described in Section 6.

5.2 CRITERIA

Any quantitative method for hazardous areas requires risk criteria of some kind, even if these are implicit rather than explicit.

The risk exists both to plant and to people. Industrial codes take account of both aspects.

For injury risk, the most usual approach is the use of a criterion of absolute risk. As far as hazardous area classification is concerned, the risk is usually taken to be primarily that to workers on the plant and this approach is followed here, but it is for consideration how far risk to the public should also be taken into account.

A criterion for risk to workers on a particular plant which has been in widespread use in the chemical industry for nearly two decades is the Fatal Accident Rate (FAR) (Kletz, 1971). The FAR is itself is defined as deaths per 10^8 exposed hours and is therefore simply another way of expressing the average annual risk to a worker.

The philosophy behind using the FAR as a risk criterion is that in designing a plant which handles hazardous materials and where there may be special hazards, the approach should be to design so that a worker on the plant is not exposed to an overall risk greater than in industry generally.

When the criterion was originally formulated, the value for the FAR both for British industry overall and for the chemical industry was about 4. This value was taken as the design 'target'. This overall target was then divided into 10 parts, or units, 5 being assigned to everyday nontechnological hazards such as falling down a staircase and 5 to the technological hazards. Then for any particular hazard one part, or unit, of the FAR is assigned as the target value, ie an FAR of 0.4 or less.

The FAR criterion is worth consideration for use in the present context. But there are certain arguments against it. One is that the risk from ignition of leaks is rather different from the other technological risks in that it occurs on so many plants. Generally, the FAR approach is used where the risk is specific to the plant concerned. Moreover, this approach does have the disadvantage that it would use up one of the five FAR units available.

More recently the HSE has been active in the development of risk criteria. Criteria have been published related to the tolerability of risk from nuclear installations (HSE, 1988) and for land use planning in the vicinity of major industrial hazards (HSE, 1989).

Whatever criterion is used, it is important that the method of assessment should be in accordance with good practice in risk assessment.

There is an alternative approach to the use of absolute risk. This is the use of a criterion of equivalent risk. In this case the proposed design is compared with an established design which has become accepted even though it may never have been justified by any absolute risk calculation. An example is the possible use of trip systems instead of pressure relief valves to protect against overpressure: the use of pressure relief valves is generally considered to give a sufficient degree of protection and it has been argued that provided they are designed for an equivalent degree of reliability trip systems should also be acceptable (Kletz, 1974). The equivalence of the risk would in this case be demonstrated by calculating the reliability of both the pressure relief valve and trip systems.

As with absolute risk, so with equivalent risk it is important that the risk assessment follow good practice, and in particular that like be compared with like.

The equivalent risk concept is already implicit in any use of the source of hazard (SoH) method in which the basic method with fixed hole size is used but the zone distance is adjusted to allow for differences in the fluid and operating conditions. In this case it is actually the hazard range which is calculated and it is assumed that the risk is effectively the same in the two cases.

6. CLASSIFICATION OF HAZARDOUS LOCATIONS: PROJECT STRATEGY

6.1 INSTALLATIONS COVERED

The types of plant considered here include:

Chemical plants

Oil refineries

Gas installations

Other process plants

Offshore oil and gas platforms

6.2 QUANTITATIVE APPROACHES

The approach to quantification adopted in this work is as follows.

First, an attempt has been made to assemble the basic information required for a quantitative approach. These include data on the inventory of leak sources on a typical plant; the national inventory of plants; the standard hole sizes used by industry; estimates of leak frequencies and hole size distributions of equipment; a set of models for emission and dispersion; physical property correlations for a set of representative flammable fluids; a numerical investigation of hazard ranges; estimates of ignition and explosion probability; and finally frequency of fire and vapour cloud explosions in plant.

Of particular importance for a quantitative approach are data on leak frequency and hole size distribution and on ignition and explosion probability. A considerable effort has been devoted to the problem of leak data.

As a means of checking the data a fire and explosion model has been constructed which allows an estimate to be made for a standard plant of the frequency of leaks, fires and vapour cloud explosions using data on the inventory of leak sources on the plant, leak frequencies, hole size distributions and ignition and explosion probabilities. The model has been used to obtain estimates of these critical parameters by successive adjustment of the initial estimates of their values. The validity of the estimates may be assessed by consideration of the degree of adjustment which has proved necessary and by the intermediate results given by the model in terms of the proportion of leaks, fires and explosions attributable to different leak sources and different fluid phases.

A strategy for using these data to explore the effect of different zoning policies has been formulated, but there has not been time within the project to implement it. The approach suggested is to explore zoning policies using a computer-based plant layout package. Information would be entered on the profile of leak sources, on the leak frequency and hole size distribution and on the travel distances from the emission and dispersion models together with initial estimates of the density of ignition sources both outside the zones defined by existing codes and also inside the zones, the latter being due to failure completely to control such sources. The initial estimates of the density of ignition sources would then be adjusted so as to give reasonable agreement with the estimates of the ignition and explosion probabilities and of the proportion of ignitions attributable to particular types of ignition source. Having thus obtained a model applicable to the present situation, the model would then be used to explore the effect of altering the zone definitions and distances.

Studies of this kind could provide the basis for general guidance, derived from quantitative work, on zoning. It is envisioned that such guidance would be the practical outcome of the work.

Work has also been done within the project on the effect on hazard distances of fluid properties, operating conditions and hole size. Sets of representative flammable fluids and of models for emission, vaporisation and dispersion have been defined. These have been used in conjunction with a set of arbitrarily defined hole sizes to explore hazard distances and to compare the hazard distances for different fluids given in different codes.

The comparative study of hazard distances may be used to formulate proposals for relating zone distances to the flammable fluid and its operating conditions. This provides the basis for a more rational approach to the setting of zone distances for one fluid relative to another, but it leaves open the question of whether the current absolute zone distances and definitions are the most appropriate.

Work of a rather tentative nature has been done on an ignition model. The purpose of this model is to explore the different types of ignition which together determine the overall ignition probability. These include ignition within a zone, ignition in a non-hazardous area and self-ignition. This model is speculative, but does draw attention to a number of key features.

A discussion of the outcome of these quantitative approaches is given in Section 22.

7.   LEAK SOURCES AND SCENARIOS

The sources of leaks on process plants considered in this work are the common sources. Process plants contain a quite large number of potential leak sources, but the list given below is believed to cover those which need to be considered here.

The zone distances generally used are intended to prevent ignition of typical leaks from these leak sources but they are not sufficient to ensure this for catastrophic leaks.

7.1   EQUIPMENT SUBJECT TO LEAKS
The principal types of equipment which are subject to leaks are those given in Table 7.1.

Table 7.1   Principal types of equipment subject to leaks

Pipework
   Flanges
Small bore connections
Instrument connections
Hoses
Valves
Pumps
Compressors
Agitators
Drain points
Sample points
Open ended lines
Atmospheric vents
Pressure relief valves
Drains
Loading arms

Note:
Releases also arise from:
Operational activities
Maintenance activities
Transport activities

7.2 TYPES OF LEAK ON EQUIPMENT

The principal types of leak from the leak sources on the equipment listed in Table 7.1 are given in Table 7.2.

Table 7.2 Some principal types of leak on equipment

Equipment	Type of leak
Pipework	Full bore (guillotine) rupture
	Pipe splits
Flanges	Gasket failures
Small bore connections	Full bore rupture
Instrument connections	Full bore rupture
Hoses	Full bore rupture
Valves	Valve rupture
	Stem leak
Pumps	Pump rupture
	Seal leak
Centrigual compressors	Seal leak
Reciprocating compressors	Seal leak
	Valve chamber release
Agitators	Seal leak
Drain/sample points	Full bore release
Open ended lines	Full bore release
Atmospheric vents	Full bore release
Pressure relief valves	Full bore release
Drains	Rising vapours
Loading arms	Full bore release

7.3   LEAK SCENARIOS

There are certain standard scenarios which may be defined for the emission and dispersion of flammable materials escaping from the plant.

The principal features which determine these scenarios are:

1.   Fluid
(a)   gas/vapour
(b)   liquid
2.   Fluid momentum
(a)   high
(b)   low
3.   Fluid state (for liquid)
(a)   subcooled (non-flashing)
(b)   superheated (flashing)
4.   Fluid density
(a)   lighter than air
(b)   neutral density
(c)   heavier than air
5.   Leak duration
(a)   instantaneous
(b)   quasi-instantaneous
(c)   continuous.

The corresponding leak scenarios are given in Table 7.3.

Table 7.3   General scenarios for leaks from equipment

Fluid	*Fluid momentum*	
Gas/vapour	High – jet	
	Low – plume	
Liquid	High – jet	Disintegrating liquid jet[1]
	Low – trickle	Evaporating pool[2]

Note:

(1) If the liquid flashes, there will be a flashing, disintegrating jet and an evaporating trail of liquid on the ground. If the liquid does not flash, there will be simply a disintegrating jet and an evaporating trail.
(2) If the liquid flashes, there will in principle be a flashing trickle before the liquid reaches the ground, but this flashoff is here lumped with the evaporation from the pool.

7.4   LEAK SOURCES vs SCENARIOS

The leak scenarios which go with particular leak sources ar   iown in Table 7.4.

These leak scenarios provide the basis for modelling the releases, as described in Section 12.

Table 7.4   Leak scenarios for particular leak sources

Equipment	Leak scenario
A Gas/vapour only	
Centrifugal compressors: seal leak	Gas jet
Reciprocating compressors: seal leak	Gas jet
Agitator: seal leak	Gas/vapour jet
Drains	Plume
B Liquid only	
Hose	Disintegrating liquid jet, or evaporating pool
Pump	As hose
Drain/sample point	As hose
C Gas or liquid	
Pipework including flanges	Gas jet, disintegrating liquid jet, or evaporating pool
Small bore connections	As pipework
Instrument connections	As pipework
Valves	As pipework
Loading arms	As pipework
D Gas/vapour usually, liquid possible	
Atmospheric vents	Gas plume, or liquid trickle
Pressure relief valve	Gas jet, or disintegrating liquid jet

8. INVENTORY OF LEAK SOURCES

8.1 PRINCIPAL LEAK SOURCES

The determination of the risk from flammable leaks on process plant requires information on the number of leak sources on such plants.

The principal leak sources considered were given in Table 7.1.

8.2 PLANT PROFILES FROM FUGITIVE EMISSION STUDIES

There are available several papers giving for typical plants the number of each of the principal types of leak source. Most of these are concerned with fugitive emissions.

A review of information on plant inventories of leak sources is given in Appendix 5. From these profiles it is possible to make estimates of the ratios of particular types of equipment and to build up a plant profile.

8.3 PLANT PROFILES FROM MATERIAL RUNOFFS

Further information is available from material runoffs and this also is described in Appendix 5. An account of the pipework and fittings size distribution for a set of plants has been given by Hooper (1982).

In addition, a study has also been made by the authors of the inventory of certain principal items, namely pipe lengths and number of flanges, valves and pumps on two major plants, one a medium sized chemical plant and one an oil production platform.

8.4 PROFILES FOR 'STANDARD' PLANTS

From the above data a profile of the inventory of these items on a 'standard' plant has been derived as given in Table 8.1.

The plant profile given in Table 8.1 gives a rather large number of pipe sizes. In order to obtain a more manageable profile this has been converted into the profile of an 'equivalent standard plant' in which the pipe sizes are limited to diameters of 25, 50, 100 and 300 mm. This profile is shown in Table 8.2. The basis of the conversion is that both plants have approximately equal quantities of small, medium and large pipework. The proportions of the pipework carrying gas, non-flashing liquid, flashing liquid are estimates made from examination of the data given with the material runoffs.

Table 8.1 Inventory of leak sources: standard plant

A Pipework, flanged joints, valves

Pipe diameter (mm)	Pipe length (m)	(%)	Flanged joints	Valves	Proportion on gas (%)
15	525	3.5	100	120	0
20	90	0.6	200	100	0
25	3000	20.0	600	500	0
40	2100	14.0	420	200	5
50	2100	14.0	650	300	8
80	3300	22.0	530	140	8
100	1050	7.0	160	60	8
150	1050	7.0	120	40	8
200	600	4.0	100	18	10
250	300	2.0	55	10	20
300	150	1.0	30	5	20
350	45	0.3	5	2	20
400	525	3.5	20	3	20
450	45	0.3	5	–	20
500	120	0.8	5	2	20
Total	15000		3000	1500	

B Pumps

No. of pumps = 25

Table 8.2   Inventory of leak sources: equivalent standard plant

A Pipework, flanged joints, valves

Pipe diameter	Pipe length		Flanged joints	Valves	Proportion on		
					Liquid	Gas	Flashing liquid
(mm)	(m)	(%)			(%)	(%)	(%)
25	3750	25	900	720	50	0	50
50	4200	28	1070	500	40	15	45
100	5400	36	810	240	40	15	45
300	1650	11	220	40	30	25	40
Total	15000		3000	1500			

B Pumps

Pipe diameter (mm)	Liquid	No. of pumps Flashing liquid	Total
50	11	4	15
100	6	4	10
Total	17	8	25

C Small bore connections

Pipe diameter (mm)	Liquid	No. of connections Gas	Flashing liquid	Total
10	180	68	202	450

9. INVENTORY OF MAJOR PLANTS

The determination of the frequency of incidents due to flammable leaks, and particularly the frequency of more serious events such as fires and vapour cloud explosions, requires information on the number of plants at risk.

Estimates of the number of major plants handling flammables are given in Appendix 6. The basis of the estimates is as follows. For major chemical and petrochemical plants the numbers are based on figures given in the Directories published by SRI International. For refineries they are based on data in the International Petroleum Encyclopaedia supplemented by surveys in the *Oil and Gas Journal*. The number of major units per refinery is taken as 5.

To these are added LPG storage installations and natural gas installations. The numbers of these are based on the numbers in the UK notified under the NIHHS Regulations, the numbers for the US and Western Europe (WE) being scaled up in proportion to the number of chemical/petrochemical plants and refinery units in those areas.

The number of major plants handling flammables is estimated to be as follows:

No. of major plants handling flammables in US = 10867
No. of major plants handling flammables in WE = 8775
No. of major plants handling flammables in UK = 1110

The figure for the UK is made up of 260 chemical and petrochemical and refinery units, 450 LPG installations and 400 natural gas units.

However, as indicated in Appendix 6, the bulk of these plants are LPG and natural gas installations, which are apparently less prone to fires and vapour cloud explosions. The number of plants of the refinery unit type is there given as follows:

No. of refinery unit type plants in UK = 180
No. of refinery unit type plants in US = 1760
No. of refinery unit type plants in WE = 1422

It is emphasised that these estimates have been derived for the specific purpose of crosschecking estimates of the frequency of fires and vapour cloud explosions on such plants.

10. EQUIPMENT HOLE SIZE DISTRIBUTION: ESTIMATES BASED O HAZARDOUS AREA CLASSIFICATION

The estimation of hole size is the most difficult problem encountered in the present work. It is not usually the practice to keep records of leaks as such and it is difficult to deduce hole size from records for repairs.

Methods used in industrial codes or in-house practices for determining zone dimensions generally utilise hole sizes which are based partly on engineering considerations and partly on expert judgment. For example, the size of the more frequent leaks from a flange may be taken as some arbitrary value, while the size of a less frequent, but larger, leak may be taken as that equivalent to the loss of a section of the gasket between bolt holes.

It was considered useful to collect information on the hole sizes used in industrial work on hazardous area classification and to derive from these a set of standard hole sizes. These hole sizes may be used in investigations of the effect of fluid properties and operating conditions on hazard ranges.

The approach adopted is as follows. A set of hole sizes, or classes, has been defined, applicable to all types of equipment, which covers the sizes used in industrial codes and practices.

Then for a given type of equipment one or more of these hole classes has been assigned. The assignment is based on the sizes used for this equipment in the industrial codes and practices.

For some items only one hole size has been adopted. For others several values are applicable. Thus for flanges there are separate sizes for leaks due to complete loss of a section and for other leaks. For valves there are separate sizes for normal and for severe duty or large valves. For centrifugal pumps and compressors there is a range of sizes, depending on the type of seal and on the shaft diameter.

While the foregoing approach is in line with that generally adopted in hazardous area classification work, it is rather arbitrary. An attempt has therefore been made to obtain estimates of leak frequency and hole size from the literature on hazard assessment which can be cross-checked with historical data on fire and vapour cloud explosions on plant. For this purpose a fire and explosion model has been developed. This model is described in Section 11 and an investigation using this model in Section 18.

For those cases where there are several hole sizes, an attempt was originally made to determine a hole size distribution, sometimes known as a severity fraction. This aspect was not pursued, since it was superseded by the fire and explosion study.

Although standard hole sizes are arbitrary, there is nevertheless considerable value in their use as practised in industry. It permits estimates to be made of the hazard range for the more common leaks and credit to be obtained for the use of improved mechanical engineering.

The approximate correspondence between the standard hole sizes defined here and those used in the fire and explosion study is given in Section 11.

10.1 STANDARD HOLE SIZE CLASSES

The authors have had access to a number of in-house methods for zone sizing. The hole sizes given in these methods range from 0.25 mm^2 or less up to about 250 mm^2.

The approach adopted here is based on the following set of hole sizes, or classes, given in Table 10.1. This set of hole classes covers the range of sizes used in industry codes and practices and the interval between the individual sizes is such that there is available a hole class which corresponds reasonably closely to the individual hole sizes used by industry.

Table 10.1   Standard hole size classes used in this work

Class	Hole area (mm^2)
1	0.1
2	0.25
3	0.5
4	1.0
5	2.5
6	5.0
7	10
8	25
9	50
10	100
11	175
12	250

10.2   STANDARD HOLE SIZE CLASSES FOR EQUIPMENT

The hole sizes commonly used in industrial codes and practices for some principal types of equipment are as follows.

The leak sources considered are:

Flanges
Valves
Pumps
Reciprocating compressors
Centrifugal compressors
Agitators
Small bore connections
Drain and sample points

The standard classes of hole size assigned to these items are given in Appendix 7.

It should be noted that from the information obtained, it appears to be practice to assign to some items of equipment several hole classes but to others only one class.

11. EQUIPMENT LEAK FREQUENCY AND HOLE SIZE DISTRIBUTION: ESTIMATES BASED ON HAZARD ASSESSMENT

The classical approach to hazard assessment involves the estimation of the frequency of failure of equipment, of the proportion of failures which are leaks and of the size distribution of these leaks. An attempt has been made to obtain these estimates, mainly from the literature.

11.1 LEAK FREQUENCY AND HOLE SIZE DISTRIBUTION
An review has been made of the failure and leak frequency of equipment. This review is described in Appendix 8.

The principal items of equipment considered in the review are pipework, flanges, valves, pumps and small bore connections. These items contribute a large proportion of the leaks which lead to fires and vapour cloud explosions on plants.

The number of leak sizes used in this work is three for pipework, valves and pumps and two for flanges and small bore connections. This contrasts with the number of leak sizes used in the previous section, which was based on industrial practice.

11.2 LEAK FREQUENCY AND HOLE SIZE DISTRIBUTION: ADOPTED VALUES
From the estimates given in Appendix 8 values of leak frequency and leak size have been adopted for use in conjunction with the equivalent standard plant described in Section 8. These values are shown in Table 11.1.

The estimates given in Table 11.1 have been used as the initial values in the numerical investigation of fires and vapour cloud explosions in Section 18.

An approximate correspondence (factor of about 2–3) between the standard hole sizes and the hole sizes used in the fire and explosion model is shown in Table 11.2.

Table 11.1 Leak frequency and size: adopted values

A Pipework

	Leak frequency (leaks/m y)			
Pipe diameter (m)	*0.025*	*0.050*	*0.100*	*0.300*
Rupture leak	10^{-6}	10^{-6}	3×10^{-7}	10^{-7}
Major leak	10^{-5}	10^{-5}	6×10^{-6}	3×10^{-6}
Minor leak	10^{-4}	10^{-4}	3×10^{-5}	10^{-5}

B Flanges

	Leak frequency (leaks/y)
Pipe diameter (m)	*All sizes*
Section leak	10^{-4}
Minor leak	10^{-3}

C Valves

	Leak frequency (leaks/y)
Pipe diameter (m)	*All sizes*
Rupture leaks	10^{-5}
Major leaks	10^{-4}
Minor leaks	10^{-3}

D Pumps

	Leak frequency (leaks/y)
Pipe diameter (m)	*All sizes*
Rupture leak	3×10^{-5}
Major leak	3×10^{-4}
Minor leak	3×10^{-3}

E Small bore connections

	Leak frequency (leaks/y)
Pipe diameter (m)	*0.016*
Rupture leak	5×10^{-4}
Major leak	5×10^{-3}

Notes:

(a) For pipework, valves and pumps the definitions of hole sizes are

Rupture leak area = A
Major leak area = 0.1A
Minor leak area = 0.01A

where A is cross-sectional area of pipe.

(b) For flanges the definition of hole size is

Section leak area = A
Minor leak area = 0.1A

where A is the cross-sectional area of the hole defined by the part of the circumference between adjacent bolts and by the thickness of the gasket.

Table 11.2   Approximate correspondence between standard hole sizes and hole sizes used in fire and explosion model

Equipment

	Standard hole sizes	*Fire and explosion model hole sizes*
Flange	CAF section leak	Major leak
	Smaller leak	Minor leak
Valve	Leak on severe duty	Minor leak
Pump	Leak from pump with mechanical seal, no throttle bush	Minor leak
Small bore connection	Leak	< < Major leak

12. MODELS OF EMISSION AND DISPERSION

A system of models for emission and dispersion of flammable gases and vapours on process plant is required. These models are described here.

12.1 LITERATURE MODEL SYSTEMS
There already exist in the literature several model systems which are relevant to this problem. They include those of:

1. O'Shea
2. BS 5345 Draft Appendix A
3. Mecklenburgh
4. Powell.

A set of models has been given by O'Shea (1982) and another in the draft Appendix A to BS 5345. The two sets of models are very similar. Another set of models is given in *Process Plant Layout* by Mecklenburgh (1985). Also relevant is the set of models presented by Powell (1984) in a paper on fugitive emissions.

A summary of these model systems is given in Table 12.1.

12.2 SPECIFICATION OF A MODEL SYSTEM
The model system required for the present work has been specified as follows:

EMISSION
Gas/vapour flow
    Subsonic
    Sonic
Liquid flow
Flashing flow
Coefficient of discharge

Table 12.1 Some systems of models given in the literature

		O'Shea	BS 5345	Mecklen-burgh	Powell
A.	*Fluid emission*				
A1.	Liquid flow	Y	Y	Y	Y
A2.	Gas/vapour flow:				
	a) Subsonic	Y	Y	Y	
	b) Sonic	Y	Y	Y	
A3.	Flashing flow	Y	Y	Y	
A4.	Subsidiary data				
	a) Coefficient	A1-A3		A1-A3	
	b) Diameter			A1-A3	
	c) Density	A3		A1-A3	
B.	*Flashing liquids*				
B1.	Flash fraction	Y	Y	Y	Y
B2.	Spray fraction			Y	
C.	*Evaporation of volatile liquid*				
C1	Evaporation rate	Y	Y	Y	Y
C2	Pool size			Y	
D.	*Evaporation of cryogenic liquid*				
D1	Evaporation rate			Y	
E.	*Liquid dispersion as jet (jet throw)*				
E1	Jet range (theoretical)	Y	Y		
E2	Jet range (practical)	Y			
E3	Subsidiary data a) Coefficient		Y		
F.	*Gas dispersion as jet*				
F1	Jet range	Y	Y	Y	
F2	Subsidiary data				
	a) Effective diameter	Y	Y	Y	
	b) Effective velocity		Y	Y	
	c) Effective density		Y		
G.	*Gas dispersion as passive cloud*				
G1	Point source, ground level	Y	Y	Y	Y
G2	Point source, high level		Y		
G3	Line source, ground level		Y		

Sources: O'Shea (1982); Mecklenburgh (1985); Powell (1984)

FLASHING LIQUIDS
Flash fraction
Spray fraction

VAPORISATION OF LIQUIDS
Pool size
Vaporisation of volatile liquid
Vaporisation of cryogenic liquid

DISPERSION
Gas/vapour jets
    Subsonic
    Sonic
Liquid jets
Gas dispersion (passive)

This system of models does not cover all the situations which may arise. It is not difficult to envision cases for which the model system given may be inadequate. For example, it is restricted to single components. It is considered, however, that it is adequate for the purpose of estimating hazard ranges for the types of release of principal interest in the present work, bearing in mind the uncertainties in the other features such as hole size.

12.3 DESCRIPTION OF THE MODEL SYSTEM

The principal models used are as follows. The models are given here without further comment. Commentary on the models is given in Appendix 9.

EMISSION
Gas/vapour
For subsonic flow

$$G = \frac{C_D}{v_2}\sqrt{\left\{2P_1 v_1 \frac{k}{k-1}\left[1 - \left(\frac{P_2}{P_1}\right)^{\left(\frac{k-1}{k}\right)}\right]\right\}}$$

(12.1)

where C_D is the coefficient of discharge, G the mass flow per unit area, or mass velocity (kg/m^{2}s), P the absolute pressure (N/m^2), v the specific volume of the gas (m^3/kg), k the expansion index and subscripts 1 and 2 are initial and final.

The criterion for sonic flow is

$$\eta_c = \left(\frac{2}{k+1}\right)^{\left(\frac{k}{k-1}\right)}$$ (12.2)

with

$$\eta = \frac{P_2}{P_1}$$ (12.3)

where η is a pressure ratio and subscript c is critical.

For sonic flow

$$G = C_D\sqrt{\frac{P_1 k}{v_1}\left(\frac{2}{k+1}\right)^{\left(\frac{k+1}{k-1}\right)}}$$ (12.4)

Liquid

$$G = C_D\sqrt{2\rho_L(P_1 - P_2)}$$ (12.5)

where ρ_L is the density of the liquid (kg/m^3).

Coefficient of discharge
The coefficient of discharge is

$C_D = 0.62$	Orifices	(12.6a)
$= 0.82$	External mouthpiece	(12.6b)
$= 1/(1 + K)^{0.5}$	General (see text)	(12.6c)

with

$$K = K_e + K_f + K_{fi}$$ (12.7)

where K is the total number of velocity heads lost, K_e the number of velocity heads lost at the entrance, K_f the number velocity heads lost due to friction and K_{fi} the number of velocity heads lost due to fittings.

Equation (12.6b) is used for an external mouthpiece, nozzle or pipe stub and Equation (12.6c) for somewhat longer, but still short, sections of pipe.

Two-phase flashing flow

$$G = C_D f(l) \sqrt{(P_1)} \qquad (12.8)$$

where l is the length of pipe (m) and $f(l)$ is a function of pipe length as given in Figure 12.1.

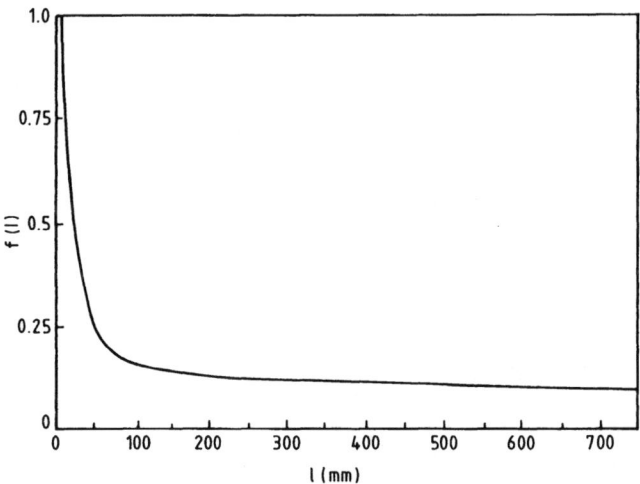

Figure 12.1   Flow correction factor f(1) for two-phase flow in a short pipe section (after Fletcher and Johnson, 1984)

FLASHING LIQUIDS
Flash fraction

$$\varphi_f = \frac{c_{pL}}{\Delta H_v}(T_1 - T_b) \qquad (12.9)$$

where c_{pL} is the specific heat of the liquid (kJ/kg K), T the absolute temperature (K), ΔH_v the latent heat of vaporisation (kJ/kg), φ_f the mass fraction flashing off as vapour and subscript b is normal boiling point.

45

Spray fraction

$$\varphi_s = \varphi \qquad\qquad \varphi \leqslant 0.5 \qquad\qquad (12.10a)$$

$$\varphi_s = 1 - \varphi \qquad\qquad \varphi > 0.5 \qquad\qquad (12.10b)$$

$$\varphi_t = \varphi_f + \varphi_s \qquad\qquad (12.11)$$

where φ_s the mass fraction forming spray and φ_t the mass fraction of vapour and spray.

VAPORISATION OF LIQUIDS
Pool area
For continuous leaks

$$A_p = \frac{\pi}{4}\left(\frac{2048 V t^3}{81}\right)^{1/2} \qquad\qquad (12.12a)$$

or

$$A_p = \frac{Vt}{h_m} \qquad\qquad (12.12b)$$

where A_p is the area of the pool (m^2), h_m the depth of the pool (m), t time (s) and V the volumetric flow (m^3/s).

Values of the depth h_m are

$$h_m = 5 \times 10^{-3} \quad \text{Concrete}$$

$$= 10 \times 10^{-3} \quad \text{Gravel}$$

The equation to be used, equation (12.12a) or (12.12b), is that which gives the smallest pool area.

Vaporisation of volatile liquid
The vaporisation of a continuous spill of a volatile liquid before significant chilling occurs is given by

46

$$G = \lambda u \rho_v \frac{P_v}{P_a} \left(\frac{2}{u^2 D_p} \right)^{n_c} \tag{12.13}$$

with

$$n_c = \frac{n}{2 + n} \tag{12.14}$$

where P_a is atmospheric pressure (bar), P_v the vapour pressure of the liquid (kN/m^2), D_p the diameter of the pool (m), n Sutton's index, n_c a vaporisation index, u the wind speed (m/s), ρ_v the density of the vapour (kg/m^3) and λ a constant.

The constant λ is a function of the wind speed and the stability. In terms of the Pasquill stability categories and for SI units the values of λ and n_c are

Pasquill stability	Sutton index	n_c	λ
A	0.17	0.078	1.0×10^{-3}
B	0.20	0.091	1.2×10^{-3}
C	0.25	0.111	1.5×10^{-3}
D	0.30	0.130	1.7×10^{-3}
E	0.35	0.149	1.8×10^{-3}
F	0.44	0.180	1.8×10^{-3}

Vaporisation of a cryogenic liquid
The vaporisation of a cryogenic liquid immediately following the leak is given by

$$G = \frac{2Xk}{\Delta H_v} \left(\frac{1}{\pi \alpha} \right)^{1/2} (T_g - T_L) t^{1/2} \tag{12.15}$$

where k is the thermal conductivity of the substrate (kW/m K), T_g the temperature of the ground (K), T_L the temperature of the liquid (K), X the surface correction factor and α the thermal diffusivity of the substrate (m^2/s).

Values of the surface correction factor are

$X = 1$ Non-porous ground
$X = 3$ Porous ground

DISPERSION
Gas jet
For a subsonic circular jet the concentration is given by

$$\frac{c_{cl}}{c_o} = 5\left(\frac{\rho_\alpha}{\rho_o}\right)^{0.5}\frac{d_o}{x} \tag{12.16}$$

where c is the volumetric concentration (m^3/m^3), d_o the diameter of the outlet (m), x the distance along the centreline (m), ρ_α the density of air (kg/m^3), ρ_o the density of gas at the outlet (kg/m^3) and subscript cl is centreline and o outlet.

For a sonic, or underexpanded, jet the concentration is given by

$$\frac{m_{cl}}{m_e} = 1 - \exp\left(\frac{-1}{k'\left(\frac{\rho_\alpha}{\rho_{eq}}\right)^{0.5}\frac{z^*}{r_{eq}} - X_c}\right) \tag{12.17}$$

with

$$z^* = z - 2z_b \tag{12.18}$$

$$z_b = 0.77d_e + 0.068d_e^{1.35}N \tag{12.19}$$

$$N = \frac{P_1}{P_e} \tag{12.20}$$

$$\rho_{eq} = \rho_e\frac{P_a}{P_e} \tag{12.21}$$

$$P_e = P_1\left(\frac{2}{\gamma + 1}\right)^{\left(\frac{\gamma}{\gamma - 1}\right)} \tag{12.22}$$

$$d_{eq} = d_e \left(\frac{C_D P_e}{P_\alpha} \right)^{0.5}$$ (12.23)

where d_e is the diameter of the exit pipe (m or mm), d_{eq} the equivalent diameter (m), m the mass fraction (kg/kg), N a pressure ratio, P_a atmospheric pressure (N/m^2), P_e the exit pressure (N/m^2), r_{eq} the equivalent radius (m), z the distance along the axis of the jet (m), z_b the barrel length (mm), z^* a modified axial distance (m), γ the ratio of specific heats of the gas, ρ_a the density of the ambient air (kg/m^3), ρ_e the density of the gas at the exit (kg/m^3), ρ_{eq} the density of the gas under the equivalent conditions (kg/m^3), k' and X_c are constants and subscript e is exit and eq equivalent. The dimensions are in metres except that in equation (12.19) they are in millimetres. The values of the constants k' and X_c are 0.104 and 0.7, respectively. The coefficient of discharge C_D is 0.85.

Liquid jet

$$x_t = \frac{u^2}{2g} \sin 2\alpha$$ (12.24)

$$x = 0.5 x_t$$ (12.25)

where u is the exit velocity of the jet (m/s), x the actual distance travelled by the jet (m), x_t the theoretical distance travelled by the jet in the absence of air drag and disintegration (m) and α the angle between the initial axis of the jet and the horizontal (deg).

Gas dispersion

$$\chi = \frac{Q}{\pi \sigma_y \sigma_z u}$$ (12.26)

with

$$\sigma = \exp[a + b \ln x + c(\ln x)^2]$$ (12.27)

where x, y and z are the distances in the downwind, crosswind and vertical directions (m), Q is the mass rate of release (kg/s), u the wind speed (m/s),

49

σ the dispersion coefficient (m), χ the concentration (kg/m^3), a, b and c are constants, and subscripts y and z are y and z directions.

Values of the constants a, b and c are:

				Parameter values		
		σ_y			σ_z	
Pasquill stability	a	b	c	a	b	c
A	5.357	0.8828	−0.0076	6.035	2.1097	0.2770
B	5.058	0.9024	−0.0096	4.694	1.0629	0.0136
C	4.651	0.9181	−0.0076	4.110	0.9201	−0.0020
D	4.230	0.9222	−0.0087	3.414	0.7371	−0.0316
E	3.922	0.9222	−0.0064	3.057	0.6794	−0.0450
F	3.533	0.9181	−0.0070	2.621	0.6564	−0.0540

Peak-mean concentration ratios

	Peak/mean concentration ratio
Gas jet	1.5
Passive plume	2

12.4 APPLICABILITY OF CERTAIN MODELS

The dispersion models just given are normally used over ranges appreciably greater than those of interest in hazardous area classification work. Their applicability over short ranges needs justification.

Dispersion under calm or low wind conditions is ill-defined. The extent to which it can be handled by the dispersion models used here requires consideration.

These aspects are considered in Appendix 9.

12.5 MODELS FOR EMISSION FROM FLANGES

Most of the emissions scenarios considered involve emission from a more or less circular hole. For pipe flanges, however, the hole geometry is significantly different. These have therefore been given a separate treatment, which is described in Appendix 12.

NOTATION

a	constant
A_p	area of pool (m^2)
b	constant
c	volumetric concentration (m^3/m^3)
c	constant
C_D	coefficient of discharge
c_{pL}	specific heat of liquid (kJ/kg K)
d_e	diameter of exit (m) (mm in equation (12.19))
d_{eq}	equivalent diameter (m)
d_o	diameter of outlet (m)
D_p	diameter of pool (m)
g	acceleration due to gravity (m/s^2)
G	mass flow per unit area, mass velocity (kg/m^2 s)
h_m	depth of pool (m)
ΔH_v	latent heat of vaporisation (kJ/kg)
k	expansion index
k	thermal conductivity (kW/m K) (equation (12.15) only)
k'	constant
K	number of velocity heads lost
K_e	number of velocity heads lost at entrance
K_f	number of velocity heads lost by friction
K_{ft}	number of velocity heads lost by fittings
l	length of pipe (m)
m	mass concentration (kg/kg)
n	Sutton's index
n_c	evaporation index
N	pressure ratio
P	pressure (N/m^2)
P_a	atmospheric pressure (N/m^2)
P_e	pressure at exit (N/m^2)
P_v	vapour pressure (kN/m^2)
P_1	initial pressure (N/m^2)
P_2	final pressure (N/m^2)
Q	mass rate of release (kg/s)
r_{eq}	equivalent radius (m)
t	time (s)

T absolute temperature (K)

T_g temperature of ground (K)

T_L temperature of liquid (K)

u wind speed (m/s)

u exit velocity of jet (m/s) (equation (12.24) only)

v specific volume of gas (m^3/kg)

V volumetric flow (m^3/s)

x distance in downwind direction (m)

x actual distance travelled by jet (m) (equation (12.25) only)

x_t theoretical distance travelled by jet in absence of air drag and disintegration (m)

X surface correction factor

X_c constant

y distance in crosswind direction (m)

z distance in vertical direction (m)

z distance along axis (equation (12.18) only) (m)

z_b barrel length (m) (mm in equation (12.19))

z^* modified axial distance (m)

Greek letters

α thermal diffusivity (m^2/s)

α angle between initial axis of jet and horizontal (deg) (equation (12.24) only)

γ ratio of specific heats of gas

η pressure ratio

λ constant

ρ_a density of air (kg/m^3)

ρ_e density of gas at exit (kg/m^3)

ρ_{eq} density of gas at equivalent conditions (kg/m^3)

ρ_L density of liquid (kg/m^3)

ρ_o density of gas at outlet (kg/m^3)

ρ_v density of vapour (kg/m^3)

σ dispersion coefficient (m)

φ_f mass fraction flashing off as vapour, flash fraction

φ_s mass fraction forming spray, spray fraction

φ_t mass fraction forming vapour and spray

χ concentration (kg/m^3)

Subscripts

b	normal boiling point
c	critical
cl	centreline
e	exit
eq	equivalent
o	outlet
y	in crosswind direction
z	in vertical direction

13. REPRESENTATIVE FLAMMABLE FLUIDS

13.1 SELECTION OF REPRESENTATIVE FLUIDS

In order to explore the hazard ranges of materials given by the models just described, it is necessary to select a representative set of flammable fluids.

The fluids selected are as follows:

1. Hydrogen
2. Methane
3. Ethylene
4. Propane
5. Acetone
6. Methyl ethyl ketone.

Hydrogen is unique in that it has a very low molecular weight. Consequently it is a searching gas and diffuses readily from plant. It is also a highly buoyant gas. Hydrogen is handled mainly as a gas.

Methane, ethylene and propane comprise a series with molecular weights 16, 28 and 44 and thus with one component of molecular weight less than, one close to and one above that of air. The normal boiling points are 109 K, 169.3 K and 231 K. All three are handled both as gases and as liquids. Methane and ethylene vaporise as cryogenic liquids, propane as a flashing liquid.

Acetone and methyl ethyl ketone are typical solvents. Acetone has molecular weight 58 and MEK 72. The normal boiling point of acetone is 329.2 K and that of MEK 352.6 K. Both vaporise as volatile liquids.

This set of fluids does not cover all cases, but is considered to be sufficiently representative to support a numerical investigation of hazard ranges.

13.2 PHYSICAL PROPERTIES OF REPRESENTATIVE FLUIDS

Data and correlations for the determination of the physical properties of the representative fluids selected are given in Appendix 10.

14. NUMERICAL INVESTIGATION OF HAZARD RANGES

Using the standard holes sizes, the models of emission and dispersion and the representative flammable fluids given in Sections 10, 12 and 13, respectively, the corresponding hazard ranges for a representative selection of leak scenarios have been investigated numerically using a computer program written for this purpose. This investigation is described in Appendix 11.

The scenarios considered are:
1. Flange leaks
(a) gas
(b) non-flashing liquid
(c) flashing liquid
2. Compressors
3. Pumps
4. Drain/sample points
(a) non-flashing liquid
(b) flashing liquid.

For all scenarios one estimate made is the distance to the lower flammability limit (LFL). In addition, for leaks of gas forming a jet the distance to 0.67 (1/1.5) times the LFL is given, while for leaks of vapour flashing off from a liquid the distance to 0.5 (1/2) times the LFL is given.

The results of the investigation are given in Table A11.1-A11.6.

A further investigation of leaks from a pipe flange is described in Appendix 12.

15. IGNITION SOURCES AND PROBABILITIES

Information on the ignition of leaks is almost as difficult to come by as information on leaks themselves, but there is a small amount available.

The fire and explosion survey of national case histories, described in Appendix 2, provides some information. Another source is reports on incidents offshore. These two sets of data are complementary in that the former tend to be for small and the latter for larger leaks.

15.1 IGNITION SOURCES

The fire and explosion survey of national case histories gives the distribution of ignition sources for closed plant shown in Table A2.5. Based on this table the data on distribution of ignition sources given in Table 15.1 may be derived. The adjusted figures in the table are obtained by

Table 15.1 Ignition sources in fire and explosion survey of national case histories

	No. of incidents	Proportion of incidents[a] (%) Unadjusted	Adjusted
Flame: general	8	9.3	16.7
LPG fired equipment	2	2.3	4.1
Hot surfaces	10	11.6	20.8
Friction	4	4.7	8.4
Electrical	8	9.3	16.7
Hot particles	3	3.5	6.3
Static electricity	6	7.0	12.5
Smoking	-	-	-
Autoignition	7	8.1	14.5
Unknown	38	44.2	-
Total	86	100.0	100.0

Note:

(a) Adjusted figures are based on eliminating the unknown category and redistributing the known ignition sources in their original relative proportions.

eliminating the unknown category and redistributing the known ignition sources in their original relative proportions.

For offshore ignition sources have been given for fires and explosions in the Gulf of Mexico (GoM) and for Norwegian North Sea (NNS) operations in reports by workers at Det Norske Veritas (eg Sofyanis, 1981; Forsth, 1981a, b, 1983). The offshore data need to be interpreted with care because the conditions for dispersion of leaks are not the same in a module as on landbased plant.

Forsth (1983) has given the data shown in Table 15.2 for ignition sources in these two locations. The number of accidents considered was for the GoM was 326 over the period 1956–81 and for the NNS 133 over an unspecified period.

Table 15.2   Ignition sources for fires and explosions on offshore platforms in the Gulf of Mexico 1956–81 and the Norwegian North Sea (Forsth, 1983)

| | Proportion of incidents[a] (%) | | | |
| | NNS | | GoM | |
	Unadjusted	Adjusted	Unadjusted	Adjusted
Welding/cutting/ grinding	18	-	59	-
Engines and exhausts	34	50.7	4	14.3
Sparks	4	6.0	1	3.6
Electrical	16	23.9	9	32.0
Hot surfaces (other than engines and exhausts)	0	0	6	21.4
Self-ignition	1	1.5	1.5	5.4
Cigarette, lighter, match	1	1.5	1.5	5.4
Other	11	16.4	5	17.9
Unknown, not reported	17	-	12	-
Total		100.0		100.0

Note:

(a) Adjusted figures are based on eliminating the hot work and unknown categories and redistributing the known ignition sources in their original relative proportions.

Both sets of data show a large proportion of ignitions by hot work. The Norwegian reports seem to indicate that in virtually all cases the hot work was both the cause of any leak and the source of ignition, or, in other words, that the hot work was not the source of ignition for a leak due to some other cause. It has been assumed that this is so for all hot work incidents in both the British and Norwegian data and such incidents have therefore been disregarded.

Also included in Table 15.2 therefore are adjusted figures based on eliminating the hot work and unknown categories and redistributing the known ignition sources in their original relative proportions.

15.2 IGNITION PROBABILITY

For probability of ignition there are a number of expert estimates and some rather sparse data.

Estimates of the probability of ignition have been given by Kletz (1977) in the context of vapour cloud explosions. Kletz states that on polyethylene plants about one leak in 10, 000 ignites and suggests this low value is due to good jet mixing with air. The leaks are mostly very small.

He also states that on a series of plants handling a hot mixture of hydrogen and hydrocarbons at 250 bar, about one leak in 30 ignites.

He argues that the probability of ignition increases with the size of leak and suggests that for large leaks ($>$ 10 ton) the probability of ignition is greater than 1 in 10 and perhaps as high as 1 in 2.

Browning (1969) has given a set of estimates of the relative probabilities of ignition, which may be regarded as based on expert judgment. For ignition under conditions of no obvious source of ignition and with explosionproof electrical equipment he gives the following probabilities of ignition:

	Relative probability of ignition
Massive LPG release	10^{-1}
Flammable liquid with flashpoint below 110°F or with temperature above flashpoint	10^{-2}
Flammable liquid with flashpoint 110–200°F	10^{-3}

He also states that if conditions are such that there are open fires or arcing electrical equipment the above values should be multiplied by 10.

Elsewhere Browning (1980) gives a table of probabilities which are evidently absolute probabilities which includes an estimate of the probability of ignition of flammable gas-liquid spills of $10^{-2} - 10^{-1}$.

Estimates of probability of ignition are given in the two Canvey Reports (HSE 1978, 1981). The First Canvey Report gives for LNG vapour clouds

	Probability of ignition
Limited releases (some tens of tonnes)	10^{-1}
Large releases	1

The Second Canvey Report gives the following 'judgment' values for ignition on site:

Sources of ignition	Probability of ignition
'None'	0.1
Very few	0.2
Few	0.5
Many	0.9

In the Rijnmond Report (Rijnmond Public Authority, 1982) the basic approach is to identify specific sources of ignition and to assign to each a probability of igniting any vapour cloud which reaches it. The probability of ignition of the cloud is then calculated by determining which ignition sources are reached by the cloud.

However, use is also made of a probability of immediate ignition, as opposed to ignition as the cloud disperses. In the study of the Oxirane propylene storage for the 9 cases of continuous releases of between 5 and 20 kg/s the value used for the probability of immediate ignition is 0.1.

Turning to offshore, Dahl *et al.* (1983) have analysed ignition data for gas and oil blowouts, which may be regarded as massive releases. The data are

Blowout	No. of blowouts	No. of ignitions	Probability of ignition
Gas	123	35	0.3
Oil	12	1	0.08

The data cover both drilling rigs and production platforms. The overall probability of ignition of blowouts is similar for the two cases.

An attempt is now made to derive from these data estimates of the probability of ignition. Three levels of leak are defined. A leak of less than 1 kg/s is classed as minor, one of more than 50 kg/s as massive and one of intermediate value as major. A massive leak gives a release of 1 te in 20 s and of 10 te in 200 s.

A distinction is also made between leaks of gas and leaks of liquid. The probability of ignition is different for the two cases.

For massive leaks the data given by Dahl *et al.* for offshore installations provide a guide. They are broadly in line with estimates made by other workers for landbased plant. Thus the probability of ignition of 0.3 for a massive gas leak is close to Kletz' estimate for that of a large vapour cloud.

As already mentioned, Browning gives an absolute value of the probability of ignition for liquid spills as lying in the range 0.01–0.1. He also gives relative estimates of 0.001 and 0.01 for lower and higher flashpoint liquids, respectively. From this the probability of ignition of a minor liquid leak is estimated as 0.01.

The most difficult value to estimate is the probability of ignition of a minor gas leak. The probabilities of ignition given by Kletz are 10^{-4} for very small leaks and 0.03 for leaks which are probably a mixture of mainly minor but some major leaks. Based on these figures, the estimate for the probability of ignition of minor gas leaks is 0.01.

The probability of ignition for major gas leaks has been obtained by taking minor and massive leaks as being 0.5 kg/s and 100 kg/s and plotting the points just derived as shown in Figure 15.1. Then taking a major leak as the geometric mean value of 10 kg/s gives from the figure probabilities of ignition for gas and liquid of 0.07 and 0.03, respectively.

The probabilities of ignition so obtained are summarised in Table 15.3. Alternatively, for particular leaks the probability of ignition may be read from Figure 15.1.

It should be noted that according to these estimates the probability of ignition of a minor leak of liquid is similar to that of a minor leak of gas, but that the probability of ignition of a massive leak of liquid is less than that of a massive leak of gas.

It should also be emphasised that the estimates obtained are approximate.

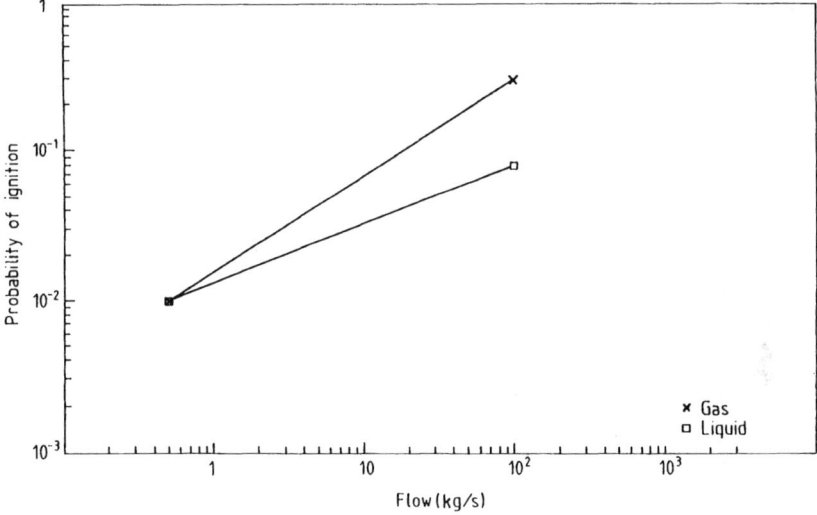

Figure 15.1   Estimated probability of ignition for leaks of gas or liquid

Table 15.3   Estimated probability of ignition of leaks of flammable gas or liquid

| Leak | Probability of ignition | |
	Gas	Liquid
Minor (< 1 kg/s)	0.01	0.01
Major (1 – 50 kg/s)	0.07	0.03
Massive (> 50 kg/s)	0.3	0.08

15.3   AN IGNITION MODEL

The foregoing estimates of the probability of ignition are based on historical data and expert judgements related to actual plants and therefore take account of the effect of current practice in hazardous area classification. They are also overall values which give no indication of the contribution of different types of ignition.

An attempt has been made to define an ignition model and to begin the construction of such a model. This work is described in Appendix 13. The model is, however, a speculative one. It has some value in drawing attention to areas which need further investigation, but it has not been used as input to other parts of the project.

16. EXPLOSION PROBABILITY

For the probability of explosion it is necessary to distinguish between the probability of explosion given ignition and the probability of explosion given a leak. The latter is the product of the probability of ignition given a leak and the probability of explosion given ignition. Unless otherwise stated the probability of explosion is used here to mean the probability of explosion given ignition.

Estimates of the probability of explosion given ignition have been made by Kletz (1977) in the context of vapour cloud explosions. He quotes the following figures:

Frequency of serious vapour cloud fires and explosions = 5/y
Frequency of serious vapour cloud explosions = 0.5/y
and derives from these for a large vapour cloud:
Probability of explosion given ignition = 0.1

Kletz also gives estimates which are evidently for the probability of explosion given leak. These are >0.1 for a large vapour cloud (10 t) and 0.001–0.01 for a medium vapour cloud (1 t or less).

Estimates of the probability of explosion given ignition are also made in the two Canvey Reports (HSE 1978, 1981). The first Canvey Report gives for a refinery:

Probability of explosion, given major fire = 0.5
Also for large LNG vapour clouds it gives:

	Probability of explosion given ignition
Large vapour clouds	1
Smaller clouds of gases other than methane	0.1
Smaller clouds of methane	0.01

Turning to offshore, the analysis of blowouts by Dahl *et al.* (1983) yields the following data:

Blowout fluid	No. of blowouts	No. of ignitions	No. of explosions	Probability of explosions given ignition
Gas	123	35	12	0.34
Oil	12	1	0	0

Sofyanos (1981) has given for fires and explosions in the Gulf of Mexico the data shown in Table 16.1.

Table 16.1 Probability of explosion given ignition for blowouts in the Gulf of Mexico (after Sofyanos, 1981)

Damage category	No. of incidents with fire and explosion	No. of incidents with explosion only	Probability of explosion given ignition
I	9	5	0.55
II	13	3	0.23
III	33	6	0.18
IV	128	22	0.18
V	143	6	0.042
Total	326	42	0.13[b]

Notes:

(a) The damage categories are in decreasing order of severity, Category I being loss of platform and Category V incident of no consequence.
(b) Weighted average.

Somewhat similar data are given in the World Offshore Accident Databank (Veritas Offshore Technology, 1988).

An attempt is now made to derive from these data estimates of the probability of explosion given ignition. The categories of leak considered are those already defined, namely minor ($<$ 1 kg/s), major (1–50 kg/s) and massive ($>$ 50 kg/s).

For massive leaks the value used is that given by Dahl *et al.* for blowouts. For minor leaks the value given by Sofyanos for Category V leaks is used.

The probability of explosion given ignition so obtained is summarised in Table 16.2.

Table 16.2   Estimated probability of explosion given ignition for leaks of flammable gas

Leak	Probability of explosion given ignition
Minor (<1 kg/s)	0.04
Major (1–50 kg/s)	0.12
Massive (>50 kg/s)	0.3

Alternatively, for particular leaks the probability of explosion given ignition may be read from Figure 16.1. The figure has been constructed by plotting the estimates given in Table 16.2, taking the leak flows for minor and massive leaks as 0.5 and 100 kg/s, respectively.

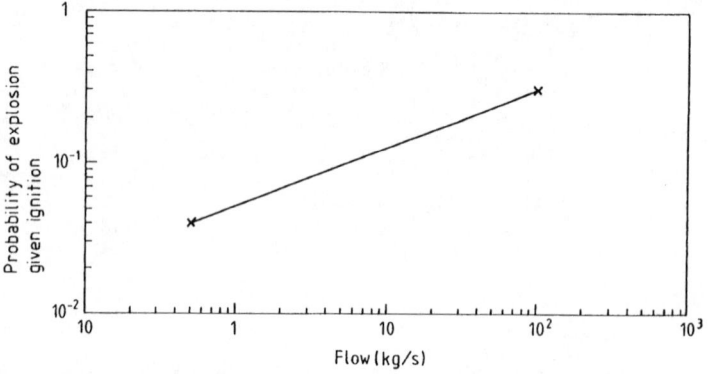

Figure 16.1   Estimated probability of explosion given ignition for leaks of gas

It is also convenient to know the probability of explosion given a leak. This probability is obtained from Table 15.3 and 16.2 and is shown in Table 16.3. The corresponding graph, constructed in the same way as Figure 16.1, is shown in Figure 16.2.

Table 16.3   Estimated probability of explosion given leak for leaks of flammable gas

Leak	Probability of ignition	Probability of explosion given ignition	Probability of explosion given leak
Minor (< 1 kg/s)	0.01	0.04	0.0004
Major (1–50 kg/s)	0.07	0.12	0.008
Massive (> 50 kg/s)	0.3	0.3	0.09

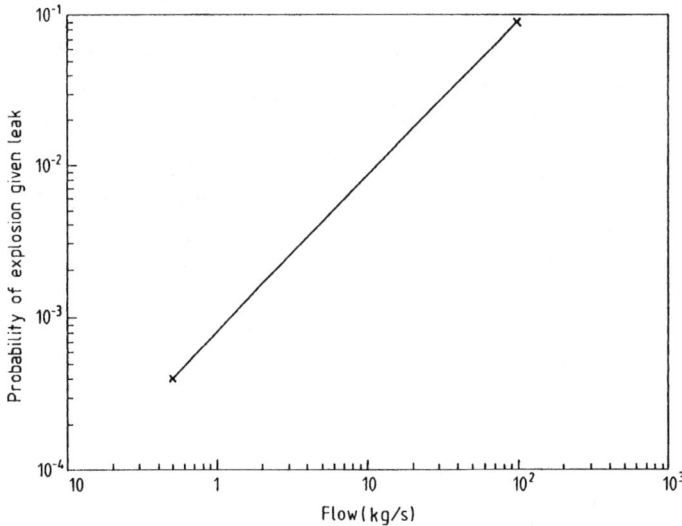

Figure 16.2   Estimated probability of explosion given leak for leaks of gas

17. FIRES AND VAPOUR CLOUD EXPLOSIONS IN PLANT

The validity of the estimates made for leak frequency and leak size distribution and for ignition and explosion probability may be cross-checked by determining the frequency of fires and explosions as derived from these estimates and comparing the results with observed values.

Estimates for the frequency of fires and vapour cloud explosions are derived in Appendix 14.

For fires an estimate has been made of the frequency of a major fire in a refinery. This is:

Frequency of major fire in a refinery = 0.1 fires/y

Then assuming 5 major units per refinery:

Frequency of major fire in a refinery unit = 0.1/5

$$= 0.02 \text{ fires/unit y}$$

For vapour cloud explosions an estimate may be made of the frequency of a vapour cloud explosion in a refinery unit type plant as described in Section 9. From the data given in Appendix 14.

Frequency of vapour cloud explosion in a refinery unit type plant

$$= 4.2 \times 10^{-4}/\text{plant y}$$

In the following section these values are compared with values synthesised from leak and ignition estimates.

18. FIRE AND EXPLOSION MODEL: NUMERICAL INVESTIGATION OF FIRES AND VAPOUR CLOUD EXPLOSIONS

The estimates made in the previous sections of the inventory of leak sources on a typical plant, the leak frequencies and leak size distributions, the emission flow rates, the probability of ignition and explosion effectively constitute the elements of a fire and explosion model.

A numerical investigation using this model has been carried out to obtain estimates of the frequencies of fires and vapour cloud explosions on process plants. These frequencies may be compared with those obtained in Section 17.

A program was written to calculate the frequency of fires and vapour cloud explosions on an equivalent standard plant using the following input data:

Inventory of equipment	Table 8.2
Leak frequency and size	Table 11.1
Ignition probability	Figure 15.1
Explosion probability, given ignition	Figure 16.2

The numerical investigation conducted using this program is described in Appendix A15.

As there described, adjustments were made to the original 'best estimates' of leak frequency and explosion probability until a set of estimates was obtained which appeared to give reasonable values not only for the overall frequencies of fires and vapour cloud explosions but also for the relative contribution of different fluid phases and individual leak sources.

Thus the leak frequencies from Table 11.1 were taken as the starting values and were successively adjusted, as were the explosion probabilities from Figure 16.1; no adjustment was found necessary for the ignition probabilities from Figure 15.1. The values of the leak frequencies which were finally used are shown in Table 18.1. The original best estimates values are shown in brackets.

Table 18.1   Fire and explosion model: values of parameters used in base case

All values to be multiplied by 10^{-4}

A Pipework

	Leak frequency (leaks/m y)			
Pipe diameter (m)	*0.025*	*0.050*	*0.100*	*0.300*
Rupture leak	0.005(0.01)	0.005(0.01)	0.0015(0.003)	0.0005(0.001)
Major leak	0.05(0.1)	0.05(0.1)	0.015(0.06)	0.005(0.03)
Minor leak	0.5(1)	0.5(1)	0.15(0.3)	0.05(0.1)

B Flanges

	Leak frequency (leaks/y)
Pipe diameter (m)	*All sizes*
Major leak	0.3(1)
Minor leak	3(10)

C Valves

	Leak frequency (leaks/y)			
Pipe diameter (m)	*0.025*	*0.050*	*0.100*	*0.300*
Rupture leak	0.01(0.1)	0.01(0.1)	0.01(0.1)	0.005(0.1)
Major leak	0.1(1)	0.1(1)	0.1(1)	0.05(1)
Minor leak	1(10)	1(10)	1(10)	0.5(10)

D Pumps

	Leak frequency (leaks/y)
Pipe diameter (m)	*All sizes*
Rupture leak	0.3(0.3)
Major leak	3(3)
Minor leak	30(30)

E Small bore connections

	Leak frequency (leaks/y)
Pipe diameter (m)	*0.01*
Rupture leak	1(5)
Major leak	10(50)

Notes:

(a) The definitions of hole sizes are as given in Table 11.1.

The hole sizes to which the leak frequencies apply are given in Table A15.1.

The ignition probabilities were not adjusted, but the explosion probabilities given ignition shown in Table 16.2, and Figure 16.1, were adjusted to:

Leak

Minor (< 1 kg/s)	0.025
Massive (> 50 kg/s)	0.25

The results of the computations for the frequency of leaks, fires and explosions per plant are given by individual fluid phase - liquid, gas or flashing liquid – in Tables A15.2-A15.4.

The frequencies of fires and explosions given in these tables are summarised by leak source and by fluid phase in Table 18.2 and 18.3.

The degree of adjustment made may be expressed as the ratio of the initial best estimate to the final adopted value. These ratios are approximately

Pipework	2
Flanges	3
Valves	10
Pumps	1
Small bore connections	5

The purpose of deriving this fire and explosion model is as follows. In order to put hazardous area classification on a quantitative basis it is very desirable to have reasonable estimates of the leak frequency and hole size distribution of the principal leak sources. The conventional approach to such a problem is to make estimates of frequency and to derive confidence limits on these estimates. For leaks the data appear to be simply too sparse to permit this approach. The problem has therefore been approached in a different way. The estimates of leak frequency and size distribution have been used to make for a typical plant estimates not only of the overall frequency of fire and explosion, but also of the frequency of leaks by leak size, by fluid phase and by leak source, and hence of the distribution of leaks by size, fluid phase and leak source. In

Table 18.2   Fire and explosion model base case: leak, fire and explosion frequencies by leak source

A Leak frequencies (leaks/plant y)

Flow (kg/s)	<1	1 – 50	>50	Total	Proportion (%)
Pipework	0.422	0.115	0.00318	0.540	21.2
Flanges	0.910	0.080	0	0.990	38.9
Valves	0.130	0.340	0.00037	0.164	6.4
Pumps	0.045	0.038	0.00030	0.083	3.3
Small bore connections	0.711	0.060	0	0.770	30.2
Total	2.22	0.33	0.0038	2.55	
Proportion (%)	87.0	12.8	0.15		

B Fire frequency (fires/plant y)

All values to be multiplied by 10^{-4}

Flow (kg/s)	<1	1 – 50	>50	Total	Proportion (%)
Pipework	42.7	31.3	5.82	80.0	25.9
Flanges	91.0	16.3	0	107	34.8
Valves	13.2	9.21	0.741	23.1	7.5
Pumps	4.50	9.02	0.504	14.0	4.5
Small bore connections	71.1	13.1	0	84.2	27.3
Total	222	79	7	308	
Proportion (%)	72.1	25.6	2.3		

C Explosion frequency (explosions/plant y)

All values to be multiplied by 10^{-4}

Flow (kg/s)	<1	1 – 50	>50	Total	Proportion (%)
Pipework	0.431	1.42	1.10	2.95	48.0
Flanges	0.753	0.418	0	1.17	19.1
Valves	0.128	0.427	0.151	0.705	11.5
Pumps	0.027	0.281	0.090	0.398	6.5
Small bore connections	0.630	0.362	0	0.991	15.0
Total	1.90	2.90	1.34	6.14	
Proportion (%)	30.9	47.3	21.8		

Table 18.3   Fire and explosion model base case: leak, fire and explosion frequencies by fluid phase

A Leak frequencies (leaks/plant y)

Flow (kg/s)	<1	1 – 50	>50	Total	Proportion (%)
Liquid	0.930	0.164	0.00184	1.10	43.0
Gas	0.285	0.00494	0.000026	0.29	11.4
Two-phase	1.00	0.157	0.00200	1.16	45.6
Total	2.22	0.326	0.0039	2.55	
Proportion (%)	87.0	12.8	0.15		

B Fire frequency (fires/plant y)

All values to be multiplied by 10^{-4}

Flow (kg/s)	<1	1 – 50	>50	Total	Proportion (%)
Liquid	93.0	32.4	1.46	127	41.1
Gas	28.6	1.7	0.077	30	9.8
Two-phase	101	44.9	5.53	151	49.1
Total	222	79	7.1	309	
Proportion (%)	72.0	25.6	2.3		

C Explosion frequency (explosions/plant y)

All values to be multiplied by 10^{-4}

Flow (kg/s)	<1	1 – 50	>50	Total	Proportion (%)
Gas	0.248	0.137	0.019	0.404	6.6
Two-phase	1.65	2.77	1.32	5.74	93.4
Total	1.90	2.90	1.34	6.14	
Proportion (%)	30.9	47.3	21.8		

other words, the model may be tested in terms not only of the final results but of a number of intermediate results. Thus for example, not only should the overall frequency of explosions check with historical values but so should the distribution of such explosions by leak source. The totality of such comparisons constitutes a fairly good crosscheck.

The following comments may be made on the results from the model for an equivalent standard plant.

The frequency of fires obtained from the model is 0.03 fires/plant y. Of these fires some 0.01 fires/plant y are from leaks of more than 1 kg/s. The model estimate needs to be corrected to allow for the fact that only a proportion of the leak sources are included in the model. In a survey of leak sources for large loss fires given by Redpath (1974) the proportion attributable to the leak sources considered in the model is some 54%, most of the rest being attributed to tanks and vessels. Allowing for the fact that these results presumably apply to all types of plant, whereas the plants of interest here are those similar to the standard plant, and that these other plants are likely to have a somewhat larger ratio of tanks and vessels to pipework, the proportion attributable to other leak sources is adjusted to 70%. Then the corrected frequency of fires obtained from the model is 0.043 fires/plant y.

This may be compared with the historical value of 0.02 fires/plant y given in Section 17. In making the comparison allowance should be made for any fire protection measures. Probably a majority of these fires would be kept by such measures below the level of a major fire. It is considered therefore that the model probably somewhat underestimates the frequency of fires. It should be noted, however, that the reliability of fixed water spray systems is not particularly high.

Turning to intermediate results, one principal crosscheck is with respect to distribution of leak sources. No data have been found which give the leak sources in sufficient detail to make this check.

The other principal crosscheck is with respect to fluid phase. Analysis of 46 incidents given by Redpath in the survey quoted indicates that some 10, or 22%, were attributable to gas. Of these 10 cases 7 involved hydrogen. For all the other materials, therefore, there were some 3 cases out of 39, or 8%. It is not clear whether any of the liquid/vapour category should be added to this. This compares with the estimated value of 9.8%.

The frequency of vapour cloud explosions obtained from the model is some 6×10^{-4} explosions/plant y. Again the model estimate needs to be corrected to allow for the fact that only a proportion of the leak sources are included in the model. From the leak sources for vapour cloud explosions given in Table 14.2 the leak sources considered in the model contribute some 68% of explosions. Then the corrected frequency of vapour cloud explosions obtained from the model is some 9×10^{-4} explosions/plant y.

This may be compared with the historical value of 4.2×10^{-4} explosions/plant y given in Section 17. In making the comparison allowance should be made for the fact that the uncorrected figures show that some 2×10^{-4} explosions/plant y are attributable to explosions of leaks of less than 1 kg/s, which after correction comes to 2.9×10^{-4} explosions/plant y. It is possible that some of these explosions are not physically possible and probable that those which may occur are under-reported. Hence it is considered that the estimate of the frequency of explosions given by the model is about right.

Turning to the intermediate results, a crosscheck may be made with respect to the distribution of leak sources. The leak sources given in Table 18.2 may be compared with those given in Table A14.2.

The other crosscheck is with respect to fluid phase. Analysis of 34 incidents given by Davenport indicates that some 11 cases out of 34, or 30%, were attributable to a fluid which was in gas or vapour form while before release. Of these 11 cases 3 involved hydrogen and 4 ethylene. For all the other materials, therefore, there were some 3 cases out of 27, or 11%. This compares with the estimated value of 6.6%.

19. VENTILATION OF INDOOR PLANTS

19.1 VENTILATION AND VENTILATION RATES

For indoor plants ventilation is of crucial importance. If the ventilation is good, a flammable atmosphere will not build up outside the envelope of the leak, whereas if it is poor, this will occur, until eventually there is a flammable atmosphere in the whole enclosure.

The general question of the ventilation of buildings is fairly well understood. It is covered by BS 5925: 1980 Code of practice for the design of buildings: ventilation principles and designing for natural ventilation.

The specific problem of ventilation of an enclosed space where a leak of flammable gas may occur has also been studied. An account of ventilation in such a situation, with particular reference to natural gas, is given in *Gas Explosions in Buildings and Heating Plant* by Harris (1983).

Ventilation is either natural draught or forced draught. Natural draught ventilation may be either thermal driven or wind driven. In the former case the movement of air is caused by difference of temperature, while in the latter it is caused by pressure differences resulting from the pressure of wind on the building. Typical natural ventilation rates are 0.5 to 3 air changes/h.

Forced draught ventilation is provided by fans. With forced draught ventilation rates can not only be higher, but can be assured.

Human comfort sets a limit to the ventilation which can be used as the normal condition. A typical rule-of-thumb is that the upper limit of the air velocity is about 0.5 m/s for comfort.

Approaches to ventilation include the following:
1. Local exhaust ventilation
2. Dilution ventilation
3. Adequate ventilation.

Local exhaust ventilation may be used to remove flammable gases or vapours where these are evolved from a well defined source either continuously or at least sufficiently often to warrant such ventilation.

The build-up of a flammable atmosphere may be prevented by the use of dilution ventilation, which involves a level of ventilation high enough to dilute below the lower flammable limit any reasonably foreseeable leak.

If this degree of ventilation cannot be achieved, it may still be possible to achieve an adequate level of ventilation, interpreted as a level sufficient to reduce the frequency of a flammable atmosphere to a low level, such as that compatible with a Zone 2 classification.

For emergency conditions it is possible to provide a higher ventilation rate, which may well exceed that compatible with comfort. There is no reason why in an emergency high ventilation rates should not be used. Such ventilation is dependent on the detection of a gas leak and hence on the reliability of the gas detectors used.

19.2   PERFECT AND IMPERFECT MIXING

If a leak of gas occurs, the part of the enclosure which is affected depends on a number of features of the leak. If the leak is a high momentum gas jet, the envelope of the flammable concentration will be relatively well defined.

If the leak has low momentum, the volume affected will depend on the density of the gas, the height of the leak source and the ventilation pattern. The normal ventilation pattern is upward flow. Harris describes experimental work which shows that with this ventilation pattern for a gas with a density lighter than air, such as methane, a high gas concentration tends to build up in the space above the leak source. This situation is illustrated in Figure 19.1. The space subject to a high gas concentration is referred to here as the gas accumulation space (GAS).

If the ventilation pattern is downward flow, which is rather unusual, this situation no longer pertains and high gas concentrations can occur below the leak source.

For a gas with a density heavier than air with upward ventilation the buoyancy and the ventilation act in opposite directions and in this case a high concentration will tend to build up in the whole space.

Thus for the normal arrangement of upward flow ventilation the gas accumulation space should be taken as the space above the leak source for a light gas and as the whole space for other gases.

For practical purposes, the gas accumulation space may be assumed to be perfectly mixed and the gas concentration in the space may

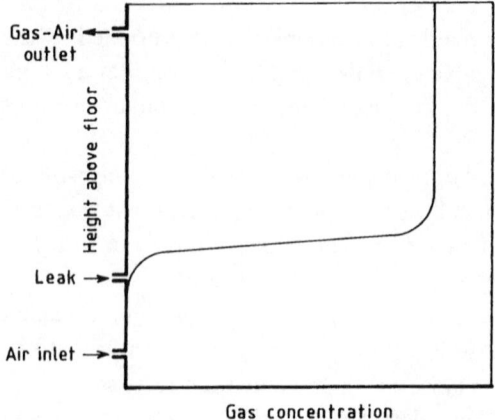

Figure 19.1 Concentration from a leak of gas lighter than air under conditions of upwards ventilation (after Harris, 1983)

be determined from the equation for a single, perfectly mixed stage, which is

$$\tau \frac{dC}{dt} = C_i - C \tag{19.1}$$

with

$$C_i = \frac{Q_g}{Q} \tag{19.2}$$

$$Q = Q_a + Q_g \tag{19.3}$$

$$\tau = \frac{V}{Q} \tag{19.4}$$

where C is the volumetric concentration in the space, C_i an effective inlet volumetric concentration defined by equation (19.2), Q the total volumetric flow rate of air and gas, Q_a the volumetric flow rate of air, Q_g the volumetric flow rate of gas, t the time, V the volume of the space and τ the volume/throughput ratio, or time constant.

Then for increase in concentration starting from the time of release

$$\frac{C}{C_i} = 1 - \exp\left(-\frac{t}{\tau}\right) \qquad (19.5)$$

and for decrease in concentration starting from cessation of the release with a concentration C_o at that time

$$\frac{C}{C_o} = \exp\left(-\frac{t}{\tau}\right) \qquad (19.6)$$

In equation (19.6) since $Q_g = 0$, the value of Q to be used is $Q = Q_a$.

19.3   BACKGROUND VENTILATION

If there is imperfect mixing of the gas, the concentration in the volume outside the zone of higher concentration may remain below the lower flammability limit provided that there is enough background, or space ventilation, to ensure that the removal of gas entering the zone of lower concentration is sufficiently rapid.

If there is perfect mixing of the gas, the concentration throughout the space may again be kept below the lower flammability limit provided the background ventilation is sufficient.

A sufficient and assured level of background ventilation is therefore essential to ensure the avoidance of a flammable concentration, whether the space is imperfectly or perfectly mixed.

19.4   INITIATION OF VENTILATION

The initiation of high ventilation rates may be effected by gas detectors. These detectors may be placed in the enclosure. Alternatively, if there is regular as well as emergency ventilation, they may be located in the ventilation duct, which should give dependable sampling. Reliance on such emergency ventilation, however, requires a high degree of reliability in the gas detectors.

20. CLASSIFICATION OF HAZARDOUS LOCATIONS: INDOOR PLANTS

On outdoor plants a flammable leak tends to disperse and its dispersion may be modelled. With indoor plants a flammable leak tends to result in a buildup of a flammable mixture, which is not readily defined and is not easy to model.

The number of fatal accidents due to flammable leaks in hazardous locations is not large. But as shown in Section 4 indoor plants appear to be the main contributor.

The classification of hazardous locations in indoor plant is also a more difficult design task. Many of the perceived problems lie in this area.

For all these reasons it is necessary to give particular consideration to indoor plants.

20.1 LEAK SOURCES ON INDOOR PLANTS

The principal leak sources on indoor plants may be classified as follows:
1. Free liquid surfaces
2. Normal leaks
a) Leaks from equipment (machinery, pipework)
b) Leaks from operations
c) Leaks from maintenance
3. Abnormal leaks
a) Leaks from equipment
b) Leaks from equipment malfunction
c) Leaks from operational errors
d) Leaks from maintenance errors.

Free surfaces of flammable liquids are not uncommon. Examples are open baths and wet solids on belts. The distance from the free surface of the flammable zone is usually very short. Factors such as limitation of worker exposure to solvents and reduction of loss of solvent will usually lead to a design where this is so.

It is normal that small leaks occur from equipment such as pumps and compressors and pipework and fittings; from operations on the plant; and from maintenance of the plant. These leaks are difficult to avoid and in some cases, such as a small continuous leak from equipment, could lead to a buildup of a flammable mixture.

Such small leaks, however, may be dealt with by ventilation. Indeed in this context 'small' may be virtually defined as the level of leak for which the ventilation solution is suitable.

A more difficult problem is posed by abnormal occurrences which lead to larger leaks. Here it is convenient to divide the leaks associated with equipment into two classes. One type is leak from equipment similar to, but generally on a larger scale than, the equipment leaks which may be regarded as normal. Again the leaks considered are those from pumps and compressors and pipework and fittings.

The other type of leak on equipment is that due to equipment malfunction. An example might be the maloperation of a machine filling bottles or aerosol cans. In such a case the actual leak would occur from a dropped bottle or damaged can rather than from the machine as such.

Larger leaks may also occur due to errors and spillages in operation and in maintenance. For such leaks, however, the question arises as to whether hazardous area classification is the proper way to deal with them or whether they are best handled by some other means such as improved overpressure protection, location of vents, design of drains, etc.

For example, if a leak occurs from a distillation plant condenser, this might be due to equipment failure such as loss of cooling water or to operator error such as turning off the cooling water supply.

This particular case highlights an important feature of leaks on indoor plants. As just stated, there are only a handful of fatal accidents due to leaks indoors. But such case histories as are available suggest that two types of leak are likely to be particularly serious. One is an emission of vapour due to a heat source such as a reboiler on a distillation column where the condenser is not operating. The other is an open drain or sample tap. Both are liable to give rise to a large release.

20.2 CONTROL OF LEAKS ON INDOOR PLANTS

This classification of leaks on indoor plants suggests certain approaches to the control of such leaks. An outline of these is given here and a more detailed discussion is given below.

Any free liquid surface should be given careful consideration and its normal conditions defined. These should be such as to ensure that the Zone 0 extends only a very short distance. Any extension of the zone should be regarded as an abnormal occurrence and dealt with as described below.

Normal leaks should be dealt with by ventilation. This is considered in Section 19.

In dealing with abnormal occurrences the first step must be to identify the potential abnormal leak sources. There would therefore seem to be a need here for a hazard identification technique specifically oriented to this. The hazard and operability (hazop) study provides a model of the type of technique required.

Control of abnormal leaks will need to draw on both preventive and protective measures. Preventive measures include both hardware such as trips and software such as operating procedures.

Protective measures also comprise both hardware and software. The former includes gas detectors and associated alarms. It also includes trips again and remote shutoff arrangements. The latter are the emergency procedures, including the procedure for re-entry into the building after evacuation on detection of a leak.

These measures will need to be supplemented by a design philosophy which recognises certain arrangements and procedures as acceptable but prohibits others. This philosophy would need to cover, for example, the carrying and transport of flammable liquids within the building.

The measures devised should be based on what is reasonably practicable. Here modern practice increasingly tends to make use of risk assessment methods. The most fruitful way of applying risk assessment to the present problem would probably be to use it to derive some general principles, thus avoiding the need for specific risk assessments, except in difficult cases.

21. CLASSIFICATION OF HAZARDOUS LOCATIONS: ZONING FOR INDOOR PLANTS

21.1 OBJECTIVES

For an indoor plant the default solution is blanket zoning. The aim of any other approach is to offer possible alternatives involving more limited zones.

It must be a prime aim of any proposals for hazardous area classification to promote in every case two essential aspects of design in such situations. These are the identification and control of any leak hazards and consideration of the alternative of locating the plant outdoors.

One of the perceived disadvantages of a blanket approach is that it makes it only too easy to neglect proper consideration of the hazards and of alternative solutions by 'playing safe'.

There are a number of features which affect the hazard of leaks indoors and any approach to their control based on zoning, and it is important to take full account of all of these.

On the assumption that an approach other than blanket zoning may be used for indoor plant, an attempt is now made to define the relevant features and to give an outline of the sort of approach which would presumably have to be taken. The scheme presented is not intended as a design approach, but rather as an attempt to make some basic distinctions. The practicalities of basing zoning on this such a scheme are discussed later.

Since an approach based on risk assessment does not yet appear feasible, use has been made of the concept that the decision on whether to classify as Zone 1 or Zone 2 should be based on the proportion of time in the year when a flammable atmosphere may be expected to be present.

21.2   OUTLINE APPROACH

The situation on an indoor plant is characterised by the following features:

RELEASE CHARACTERISTICS
1. Fluid density
(a) lighter than air
(b) heavier than' air
2. Size
3. Frequency
4. Momentum
(a) high
(b) low
5. Duration
(a) instantaneous
(b) quasi-instantaneous
(c) continuous

OTHER CHARACTERISTICS
6. Ventilation
(a) good
(b) poor
7. Restoration
(a) short period
(b) long period
8. Consequences.

Here ventilation is defined as good if it is sufficient to prevent the build-up of a flammable atmosphere (FA) outside a defined envelope near the leak source.

Restoration depends on (a) detection and (b) shutoff.

The effect of these various features may be summarised as follows:

Fluid density determines the effective volume of the space where the FA tends to build up. For a gas with a density lighter than air, such as methane, this is the volume above the level of the leak source, while for a gas with a density equal to or heavier than air it is the volume of the whole enclosure. This volume is referred to here as the gas accumulation space (GAS).

Leak size affects the size of the flammable atmosphere zone (FAZ) envelope for the good ventilation cases. If ventilation is poor, then, by definition, there is no such envelope and an FA will build up in the whole enclosure.

Leak frequency affects the zone number to be assigned.

Leak momentum determines the dispersion model to be used in the cases where there is good ventilation.

Leak duration determines the dispersion model to be used in the low momentum cases. A quasi-instantaneous release may be be treated as an instantaneous or a continous release, depending on the precise details of the release.

The implications of these statements are explored in more detail below.

It is convenient to define a hazard time as the time period in the year when an FA is present and a fractional hazard time (fht) as the proportion of the year when this is so. Define $fht_{1,2}$ as the fht criterion separating Zones 1 and 2 and $fht_{2,n}$ as the criterion separating Zone 2 and the Non-hazardous Zone.

There are six basic cases, which may be distinguished by release momentum and duration and by ventilation quality. These are as given in Table 21.1

Table 21.1 Classification of hazardous locations: base cases for indoor plants

Release momentum	Ventilation quality	Release duration	
		Instantaneous	Continuous
High	Good	Case 1	Case 1
	Poor	Case 2	Case 2
Low	Good	Case 3	Case 4
	Poor	Case 5	Case 6

These six cases are now considered in detail.

CASE 1: HIGH MOMENTUM, ALL DURATIONS, GOOD VENTILATION
Emission model: jet
Hazard time = release time
FAZ = jet envelope
FAZ number set by fht condition

CASE 2: HIGH MOMENTUM, ALL DURATIONS, POOR VENTILATION
Emission model: NA
Hazard time = time from start of release to elimination of FA in GAS.
(Actually FA will exist initially only within the jet and will take some time
to build up, so this is a simplification)
FAZ = GAS
FAZ number set by fht condition

CASE 3: LOW MOMENTUM, INSTANTANEOUS RELEASE, GOOD
VENTILATION
Emission model: puff
Hazard time = time from start of release to extinction of puff
FAZ = puff envelope
FAZ number set by fht condition

CASE 4: LOW MOMENTUM, INSTANTANEOUS RELEASE, POOR
VENTILATION
Emission model: NA
Hazard time: time from start of release to elimination of FA in GAS
FAZ = GAS
FAZ number set by fht condition

CASE 5: LOW MOMENTUM, CONTINUOUS RELEASE, GOOD
VENTILATION
Emission model: plume
Hazard time = release time
FAZ = plume envelope
FAZ number set by fht condition

CASE 6: LOW MOMENTUM, CONTINUOUS RELEASE, POOR
VENTILATION
Emission model: NA
Hazard time = time from start of release to elimination of FA in GAS
FAZ = GAS
FAZ number set by fht condition

    If the ventilation is good, there exists the prospect of having a
limited hazardous zone and avoiding the need for blanket zoning. If the
ventilation is poor, blanket zoning is the only choice.

The effect of the various features of the situation is summarised in Table 21.2.

Table 21.2   Classification of hazardous locations: role of various features in outline approach for indoor plants

Release characteristics

1. Fluid density	Definition of gas accumulation space (GAS)
2. Size	Size of flammable atmosphere zone (FAZ)
3. Frequency	Zone number
4. Momentum	Case number
5. Duration	Case number

Other characteristics

6. Ventilation	Case number
7. Restoration	Hazard time, hence zone number
8. Consequences	Fht criterion, hence zone number

21.3   PRACTICAL APPLICATION

The approach to the classification of hazardous locations for indoor plants just described is no more than an outline and is rather theoretical. There are in addition various practical aspects to be considered.

It is important that any method for indoor plants should not be applied mechanically, or naively. The features of the specific situation should be taken into account.

Some of these features are as follows:

1. Size of enclosure
2. Nature of fluids
3. Nature of plant
4. Nature of activities
5. Occupancy of enclosure.

There must be a size of enclosure below which it is not appropriate to adopt limited zoning. There are two possible bases for setting a minimum enclosure size. One is simply some absolute size. The other is a size related to the size of the estimated limited zones. It is tentatively suggested that both approaches be considered.

The nature of the fluids handled should be taken into account. If the fluid is superheated, so that an appreciable fraction will flash off on release, particular caution should be exercised.

The nature of the plant is equally important. Relevant features are the inventory which could escape, the apertures through which escape might occur and the operating conditions which determine the rate of release and the behaviour of the escaping fluid.

In some cases the situation may involve activities with flammable fluids as well as fixed plant. In particular, there may be transport of such fluids within the enclosure. It may be appropriate to seek to eliminate such activities, but in any case such a situation needs particular care.

The occupancy of the enclosure is relevant in several ways. One is that people introduce additional potential sources of ignition. Another is that if there are people in the enclosure, the probability of detection of the leak will normally be much higher. Another is that the consequences of ignition of the flammable escape are likely to be more serious.

In classification of hazardous locations the aim is to control all sources of ignition, not just electrical equipment. The zoning adopted may therefore place limitations on the activities of people in the enclosure.

21.4 MODELLING DISPERSION INDOORS

Given that limited rather than blanket zoning is to be considered, the principal cases of interest for gas are those of continuous release under conditions of good ventilation, with either high momentum (Case 1) or low momentum (Case 5). If for these cases the hole size can be estimated and the flammable envelope satisfactorily modelled, then a zone size can be defined.

As with outdoor plants, in the absence of values for the leak frequency and size distribution of equipment, it is still necessary at this stage to utilise some arbitrary hole size as the basis of design.

Assuming that this procedure is adopted, the question becomes one of estimating the flammable envelope given by the assumed hole. This may be done using appropriate models. However, the modelling of dispersion indoors is not well developed.

An essential requirement for any such modelling is that the conditions be well defined. Unless this is so, the use of models is inappropriate.

The models available for the dispersion of a high momentum gas jet are applicable indoors as well as outdoors and this case appears to present little difficulty.

The problem arises in trying to model the dispersion indoors of low momentum gas plumes. There is a large literature on the dispersion of such plumes in the atmosphere, where the dispersion is a strong function of the weather conditions, mainly wind speed and atmospheric stability.

It may be noted that in early work on gas plumes Sutton (1950) applied his model to the dispersion of a plume indoors. On the other hand the correlation of dispersion coefficients given by Pasquill (1961) is in terms of atmospheric conditions.

In considering whether dispersion models might be applicable indoors, a distinction may be made between conditions of low air speed, due to low ventilation, and of high air speed.

Experimental work outdoors has shown that at very low wind speeds, the order of 0.5 m/s, thus essentially calm conditions, dispersion is highly variable, this variability sometimes being expressed as the fit obtained to the theoretical models in terms of the Pasquill stability categories. In these terms dispersion in calm conditions has been found, as stated in Section 12, to correspond to stability categories ranging from Pasquill A to F. Indoors, however, the limit of comfort is approached at an air speed of 0.5 m/s.

In view of the above, it seems unlikely that the ventilation rate would normally be sufficiently high to allow a conventional dispersion model to be used with much confidence.

At higher air speeds, say above 2 m/s, there would seem to be a prima facie case for using such models, perhaps with neutral stability parameters, but this would need to be studied.

The foregoing comments refer primarily to dispersion of gases. As far as liquids are concerned, the spread of a liquid may be both reduced and made more predictable by the use of appropriate bunds and drains. A pool of volatile liquid may be treated as a low momentum source. The comments just made on low momentum gas releases apply here also.

A more difficult problem is presented by a liquid jet. The travel distance of such a jet is independent of hole size, at least theoretically, so that in this case uncertainty about leak size is less of a problem. However, the travel distance of a liquid jet is considerable. There seems no reason why a liquid jet model should not be used for this case, but if the liquid is at high pressure the result may well be that the predicted travel distance will equal or exceed the relevant dimension of the enclosure.

As for flashing liquids, these combine features of both gas and liquid releases. The approaches just described for low momentum gas and liquid release appear applicable in combination to a low momentum release of flashing liquid also. Similarly, a jet of flashing liquid combines the features of a liquid jet and of low momentum gas release, again often giving long travel distances.

22. CLASSIFICATION OF HAZARDOUS LOCATIONS: OVERVIEW

The problem of hazardous area classification is a complex one. The approach taken in this work has been to start with what appears to be the simplest case, that of an outdoor plant, to try to develop a quantitative approach for that case and then to consider the applicability of this basic approach to the other cases and possible modifications required for these cases.

The overall approach taken is outlined in Section 6 and is reviewed again in Section 23. In this section consideration is limited to some interim comments on the implications of the work for zoning.

22.1 OUTDOOR PLANTS

As described in Section 6, it is envisioned that in the longer term zone definitions and distances for outdoor plants might be put on a more quantitative footing by exploring the effect of different zoning policies using a computer-based plant layout package. The zoning policies would be incorporated in the model by altering the density of the various types of ignition source to represent the situation which would arise if the policies were followed, with due allowance for the failures totally to exclude ignition sources which occur in practice.

Such a study could take into account not only ignitions, but also injury to people and zoning policies could be adjusted to fit the risk criteria selected.

It is expected that such a study would serve to emphasise two important features. One is that ignition of a leak from a given source is a function not simply of the envelope around that source considered in isolation, but of the envelope created around the totality of sources. The other is that it is not only the distance between an individual leak source and an ignition source which matters but the overall density of ignition sources.

It is also expected that such an approach would highlight the fact that a given level of risk may be obtained with various combinations of Zone 1 and Zone 2 distances and that there is therefore, to some extent at least, a tradeoff between them.

Since it is not possible at this point to provide guidance derived from such a fundamental approach, it has been judged desirable to make in addition some interim comments and proposals. These relate to:

1. Definition of zones
2. Definition of hazard ranges and zone distances.

In current hazardous area classification practice Zone 0 is defined essentially as one in which a flammable atmosphere is likely to be present continuously, Zone 1 as one in which such an atmosphere is likely to occur in normal operation and Zone 2 as one in which such an atmosphere is likely to occur only as a result of some abnormal occurrence. A non-hazardous area is defined as an area not classed as Zone 0, 1 or 2.

In discussing zones, it is convenient to refer to zone distances, although strictly what matters is the volumes enclosed.

Zone 0 does not appear to present a major problem in that any space in which a flammable atmosphere is present continuously is defined by the design and it is clear that all ignition sources should be excluded from it.

The definition of Zones 1 and 2 is rather less clearcut. Attempts have been made to put the definition on a quantitative basis. One suggestion has been to define Zone 1 as a zone in which a flammable atmosphere is present for up to 1000 h/y and Zone 2 as one in which such an atmosphere is present for less than 10 h/y (but with sufficient frequency to be hazardous).

These rules-of-thumb are useful to designers in certain situations. Although this would not be its prime purpose, a fundamental study of the kind proposed could have as one outcome a refinement of these rules.

With regard to zone distances, the work done suggests that one particularly important factor affecting hazard range is the nature of the fluid. Flashing liquids are of particular importance. The leak flow rate is high because the fluid is a liquid. But a large fraction of the fluid then forms vapour or spray. Thus this case is liable to give large vapour leaks with a consequent high probability of ignition.

This is not a new result, but it is desirable that it be fully recognised in hazardous area classification. At least a relative ranking of suitable zone distances for a given type of leak source may be obtained by determining hazard ranges for different fluids, using the emission and dispersion models and, if necessary, arbitrary hole sizes.

There are appreciable differences between the zone distances given in the various codes. But, as already mentioned, there are two factors which suggest that the effect of these differences may be less than first appears. One is that for a large section of plant what matters is the envelope around that section rather than that around a particular leak source. The other is that, for large leaks particularly, the overall density of ignition sources may be as important as the distances from particular leak sources to particular ignition sources. These remarks have less bearing, of course, on isolated leak sources.

Present hazardous area classification practice involves the use of codes of various degrees of stringency. It is probably fair to say that some codes are regarded as rather conservative. This might suggest that while some lengthening of the zone distances may be desirable in the less conservative codes, particularly for flashing liquids, some relaxation might perhaps be made for the more stringent ones. However, there is some indication that the risk from ignition of flammable materials is not insignificant. It would seem preferable to defer any general revision of codes based on the approach described here until more refined models and criteria are available.

Often, however, the problems facing the designer are concerned not with the overall application of a particular code but with a relatively small number of instances where strict application of the code either creates a difficulty or appears irrational. For such situations there is, technically at least, a case for permitting limited relaxations, particularly where it can be shown by quantitative assessment to introduce only a small increment in the overall risk. It is recognised, however, that such a policy could create difficulties for regulation which may render it impractical.

22.2   INDOOR PLANTS

The hazardous area classification of enclosed plants is of particular importance. Although much of this report is concerned with the attempt to produce a quantitative method for outdoor plants, this is because such plants appear to represent a simpler, though still rather difficult, problem and their study is a necessary starting point. The problem of indoor plants is fully recognised.

The outstanding feature of an indoor plant is that flammable material from leaks does not disperse as it does on an outdoor plant. This in turn means that for an enclosed plant the following features are particularly important:

1. Special processes involving flammable materials
2. Activities involving flammable materials
3. Electrical ignition sources
4. Gas dispersion in enclosed areas
5. Ventilation of enclosed areas
6. Liquid flow in enclosed areas.

There are certain special processes such as spray booths and degreasing plants which mostly involve open surfaces and which figure highly in the accident statistics. The HSE are well aware of the hazards associated with these processes. Without a special study it is not appropriate to propose a particular zoning policy except to suggest that such a policy should be relatively stringent.

Another important source of accidents is activities, such as operations, maintenance and movement of materials. Of particular importance are activities involving handling and carrying containers with open surfaces. For a surface in a fixed item of plant the calculated zone distance is usually rather short. The problem arises from activities. In particular, these activities often result in spillage or in bringing a container of flammable liquid close to an ignition source.

Also included in activities are situations where a flammable atmosphere can occur due to some malfunction on the plant other than a leak from the basic plant envelope such as a pipe or pump leak. One example is a bottle falling from a filling machine and smashing on the floor. Another is malfunction of a heater causing vaporisation from a bath of flammable liquid.

The problem of activities is not solely, or even primarily, one of zoning. It is proposed that the principal means of dealing with it should be some form of hazard identification technique developed specifically for this purpose, perhaps some adaptation of the hazard and operability method. This aspect is discussed more fully in Section 20.

In indoor plants the problem of electrical ignition sources tends to be more prominent. In particular, it is common practice to apply blanket zoning to whole enclosed areas and to install the appropriate electric motors, switches and lighting.

In order to apply limited, as opposed to blanket, zoning to indoor plants it is necessary that the situation be reasonably defined and predictable. An attempt has been made to indicate the limits of zoning indoors in Section 21 and is therefore not repeated here.

The conclusion to be drawn from the account given there is that the designation of an area around an enclosed plant as Zone 2 and of the remaining enclosed space as non-hazardous may sometimes be acceptable practice but only subject to certain quite stringent conditions.

The source(s) of release should be well defined and activities which could give rise to releases should be absent or sufficiently well controlled that the zoning is not nullified by them. The behaviour of any release should be sufficiently predictable to allow zone distances to be set. There should be sufficient ventilation to prevent the background concentration from rising to a flammable concentration. If these conditions are fulfilled, there seems no reason why part of an enclosed space should not be designated as non-hazardous.

22.3   OFFSHORE PLATFORMS

Another hazardous area classification situation of importance is plant offshore on oil and gas production platforms.

Such platforms constitute one of the more intractable aspects of what is in any case a rather difficult problem. Although some plant is in the open, much is in modules which are partially or totally enclosed. Moreover, the plant tends to be very congested.

In view of this and of the fact that the results even for outdoor plant are provisional, it is thought better at this stage not to make proposals for offshore.

22.4   MINIMUM INVENTORY

Another aspect of the project is the question of whether it is appropriate to specify a minimum inventory below which the ignition hazard is negligible and a hazardous area classification approach is not required.

Equipment containing such a minimum inventory is likely to present a problem principally indoors and to this extent the problem may be regarded as an aspect of hazardous area classification for indoor locations.

It is clear from the frequency of accidents involving carriage of flammable liquids in open buckets, of accidents in laboratories and of fires with chip pans that even a small quantity of such liquid can present a hazard in an uncontrolled situation. Attention has already been drawn to the need to identify and control the hazards of activities in indoor plants.

The practical situation for which a minimum inventory relaxation might be helpful is presumably that involving some kind of fixed plant indoors. Following the approach adopted in the previous treatment of indoor situations of trying to define the conditions under which a limitation may be appropriate, the most important conditions in this case would seem to be that the liquid should not be under pressure, that it should not be superheated, that some form of containment such as a bund should be provided and that there should be adequate ventilation, so that if an escape occurs the liquid is contained and the amount of vapour generated is small and readily diluted by the ventilation available.

It is suggested that if a minimum inventory is to be permitted, the overall approach might be broadly on these lines. The further work proposed on ventilation and zoning of indoor plants would assist in determining the appropriate quantity.

A minimum inventory does not seem suitable for liquids held under pressure, including liquefied gases.

23. CONCLUSIONS AND RECOMMENDATIONS

The work described in this report is essentially an investigation of the feasibility of putting hazardous area classification on a more quantitative basis.

This is a difficult problem, which has exercised many people over a considerable period. The authors undertook the work with some misgivings, fearing that the results achievable for the given time and resources might be too slender to bear the weight of the expectations.

The work has not reached the point where clear and uncontentious guidance on hazardous area classification can be given. What has been done is to produce one type of overall approach, including models and risk criteria, which might form the basis of further work aimed at producing guidance.

It is not altogether a bad thing that it has not proved possible within a single project to devise a cut and dried method and that it is necessary to report interim results. There is some positive gain that these results should be available for comment by the profession before any further development of methodology takes place.

Reviews have been given of the development of hazardous area classification, current standards and codes of current performance in control of ignition sources.

There are considerable differences in some of the zone distances recommended in the various codes and examples of such differences have been given.

The risk criteria which might be used for hazardous area classification have been reviewed. One of the risk criteria widely used in industry, the Fatal Accident Rate (FAR) criterion, has been discussed. The HSE is currently developing other criteria. It has published one set of criteria for the tolerability of risk for nuclear power stations (HSE, 1988) and another for land use planning in the vicinity of major industrial hazards (HSE, 1989).

The risk to workers from ignition of flammable leaks has been reviewed. The data on which to make a judgement are sparse, but the risk does not appear to be negligible. The risk estimated for plant workers in the oil and chemical industries is an FAR of 0.56 per 10^8 working hours.

The approach taken has been to consider first outdoor plants, as being the easier case.

Basic information and models relevant to a quantitative approach have been assembled. These include inventories of leak sources on plants and the national inventory of plants; a set of standard hole sizes similar to those currently used in industry; estimates of leak frequency and hole size distribution contained from equipment failure data; a set of models for emission, vaporisation and dispersion of leaks and properties of a set of representative flammable fluids; and estimates of the probabilities of ignition and of explosion.

The main thrust of the work has been to develop concepts and methods which could form the basis of a more quantitative approach, but no attempt has been made to produce an actual design method.

The problem has been approached in two stages. The first stage is the development of a method of comparing the hazard ranges or zone distances given in different codes in respect of the factors which affect these ranges, namely the fluid properties and operating conditions, using the standard hole sizes, the emission and dispersion models and the representative flammable fluids.

The second stage is the development of a method of applying the techniques of risk assessment in order to obtain an estimate of the absolute risk implied by a particular zoning policy. This more ambitious aim is difficult to achieve, because information on leaks and ignitions is hard to obtain. What has been done is to devise an overall approach and to go part way to implementing it.

In order to make a risk assessment it is necessary to have estimates of the inventory of leak sources on a plant, the leak frequencies and hole size distributions and the ignition and explosion probabilities.

The ideal approach would be to collect data on leaks and ignitions and to present the data together with appropriate confidence limits. Unfortunately, data sufficient to support such an approach have not been found, despite considerable effort.

Accordingly, an alternative approach has been taken. The leak sources on plants have been studied and a leak source profile for a 'standard' plant has been derived. A fire and explosion model has been developed in which the inputs are the leak frequency and hole size distribution and the ignition and explosion probabilities and the outputs are the frequency of leaks, fires and vapour cloud explosions, both overall and by fluid phase and by leak sources, such as pipework, valves, etc. The outputs from the model may be compared with the corresponding historical data. The purpose of the model is to assist in obtaining improved estimates of the leak frequencies and hole size distributions and the ignition and explosion probabilities.

The way in which the model has been applied is to make initial estimates of the input values required and to adjust these to obtain a reasonable fit with historical data for the frequency of leaks, fires and vapour cloud explosions. The validity of the values so obtained may be judged partly by the degree of adjustment required and partly by the fit of the intermediate results for frequency of leaks, fires and explosions by fluid phase and by leak source.

It is envisioned that further work might proceed broadly as follows. The effect of different zoning policies would be investigated using a computer-based plant layout package. The inputs would be the layout of the standard plant, the leak frequencies and hole size distributions and the hazard ranges given by the emission and dispersion models. The zones defined by existing codes would also be an input. Estimates would then be made of the distribution of ignition sources both outside the zones and also inside them, the latter being due to failure completely to exclude such sources. These estimates would then be adjusted to give the historical ignition and explosion probabilities. Having thus obtained a model of the situation using present zoning policies, the effect of alternative policies could be explored.

These policies could include definition and range of both Zone 1 and Zone 2.

It is expected that the risk assessment approach just described would be used for the production of general guidelines along the lines of existing codes rather than for major extension of the source of hazard approach. With the latter appreciably more effort is required and uncertainty in the models and data used bulks larger.

Attention is drawn to two points which are highlighted by the work. One is the importance of self-ignition of leaks, which may account for a relatively large proportion of small leak ignitions. The other is the importance of flashing liquids such as LPG. Such a fluid combines the high emission rate of a liquid with the dispersion characteristics of a vapour and can give a long hazard range.

Concerning zoning of outdoor plants two important points may be made. One is that for a large section of plant the significant feature is the envelope around that section rather than the envelopes around individual leak sources. The other is that, for large leaks especially, the overall density of ignition sources may be as important as the distances from particular leak sources to particular ignition sources.

For outdoor plants two interim proposals on zoning are made. One is that Zone 2 distances should reflect the nature of the fluid, especially flashing liquids.

Another proposal is that the principle be considered of limited relaxations of code requirements for awkward cases where strict adhesion to the code is costly or otherwise difficult to defend and where the increment of risk due to such relaxation is estimated to be negligible.

The foregoing work applies principally to outdoor plants, but the problems of hazardous area classification of indoor plants have also been considered. Two important features of leaks in indoor plants are leaks other than those from fixed, closed plant and the various aspects of ventilation.

For indoor plants certain special processess such as spray booths and degreasing plants have been identified as being particularly prone to accidents and as needing separate treatment,

Another common source of accidents in indoor plants is activities such as movement of materials, maintenance, etc. Here attention should be directed primarily to identifying and controlling the hazards from the activity rather than to zoning as such. It is therefore proposed that a hazard identification method be developed specifically adapted to hazardous area classification.

The dispersion of leaks by ventilation is difficult to model and more work needs to be done in this area, which is of particular importance to premises other than those in the oil and chemical industries.

The application of zoning to indoor plants has been studied in the sense that an attempt has been made to define the distinctions between different situations on which a zoning approach would presumably have to be based and the conditions under which a modelling approach might be used. Essentially, the conditions are that the source of release be well defined, that any activities be sufficiently well controlled that they do not nullify the zoning, that the behaviour of any release be sufficiently predictable for zone distances to be set and that there be sufficient ventilation to prevent the background concentration from reaching a flammable concentration.

Since it is difficult to predict dispersion in congested indoor plants, these proposals do not apply to such plants.

For similar reasons, no proposals have been made concerning plant on offshore platforms.

Before guidance on hazardous area classification can be given which has a more quantitative basis, more work needs to be done. It is suggested that this work should cover the following areas:

1. Comparison of the different risk criteria available, so that the most suitable criterion can be selected and where necessary refined into a form suitable for hazardous area classification;

2. Refinement of the fire and explosion model;

3. Estimation of the probability of injury due to ignition of leaks;

4. Application of the refined fire and explosion model to zoning of outdoor plants;

5. Study of self-ignition of leaks;

6. Development of the concept of limited relaxations for outdoor plants, including criteria for deciding where such relaxations are appropriate and to what extent they should be applied;

7. Extension of the work to indoor plants. This would include:

a) development of hazard identification techniques for activities (as defined above) in indoor plant

b) examination of the mechanisms of dispersion by ventilation

c) identification of situations where quantification of zoning is applicable to indoor plants;

8. Application of quantified hazardous area classification concepts to offshore platforms.

Items 1, 5, 7 (a) and (b) and 8 may need to be separate studies.

Finally, it should be emphasised that the work described has been done to assist in the development of design methods for hazardous area classification but should not itself be used for plant design.

APPENDIX A—TERMS OF REFERENCE

A.1 TERMS OF REFERENCE

The terms of reference for this project were as follows:

The overall objective of the work is to assist the IIGCHL to suggest improvements to the current UK approach to hazardous area classification and to provide background information for those involved in contributing to codes of practice in this area,

The concept underlying the investigation is that the definition of hazardous areas may benefit from a quantitative approach. However, the problems will be approached with an open mind. It may be that quantitative methods can be developed which are applicable to individual cases. It may be that generalised rules can be devised which, whilst based in part on quantitative work, do not require quantitative calculations in individual cases. Or it may be concluded that quantification does not appear to have a significant contribution to make at this stage.

The project covers five of seven topics originally identified by the IIGCHL, namely:

1. Emission
2. Dispersion on Open Plants
3. Dispersion on Closed Plants
4. Minimum Inventory
5. Zone Definition.

The approach will be to cover the whole set of tasks in a balanced way. If intractable problems are identified, as they may well be, they will be noted and bypassed unless absolutely crucial with a view to further work later. In particular, the approach to Topics 1–4 will be governed by their role as contributions to Topic 5.

If difficult problems are identified, these will be notified to the IIGCHL as the work progresses rather than at the end so that, if desired, action can be taken.

The investigators will familiarise themselves with BS 5345 and related standards and codes on hazardous areas. They will hold discussions on the problem of hazardous areas and of code interpretation with industry, covering a wide spread of firm sizes and industry sectors. They will consult the parties such as the HSE and the Institutions with interests in this problem. They will maintain close liaison with the sponsors of the work through a Steering Committee.

1. EMISSION
The overall objective is to specify a method of estimating emission frequencies, rates and durations. The detailed objectives are to:
1. Identify potential sources of release
2. Provide estimates of aperture size and frequency/probability
3. Select methods of estimating emission rates and durations
4. Specify a standard approach to emission assessment.

The work will concentrate on releases which occur in practice and are credible and will not attempt to deal with very rare, catastrophic releases, which are not covered by the area classification.

The work will be based both on search of the literature, including standards and codes, and on discussions with industry, covering both information on holes and current design methods. It will seek to identify the types of equipment and defects which result in releases and the frequency/probability of such releases, the nature and size of the holes, the appropriate methods of modelling the hole and the flow from it, and to specify a standard approach taking into account event frequency/probability and to give numerical computations and worked samples.

2. DISPERSION ON OPEN PLANTS
The overall objective is to specify a method of estimating the hazardous area resulting from dispersion, giving a method of handling emission. The detailed objectives are to:
1. Specify a set of emission scenarios and associated frequency/probability
2. Select methods of estimating vaporisation from pools (for cases where leak is a liquid)
3. Select methods of estimating dispersion of gas
4. Devise methods of defining hazardous area of gas cloud
5. Specify a standard approach to hazardous area on an open plant.

The work will again be based on literature search and discussions with industry. It will specify a comprehensive set of initial scenarios, will seek to select methods of estimating vaporisation and dispersion, to devise methods of defining the hazardous area of the cloud, to specify a standard approach and to give numerical computations and worked samples.

The dispersion methods will cover dispersion of gas of neutral, positive and negative buoyancies and dispersion over short ranges.

The method of defining the hazardous area will be based on gas dispersion, including peak-to-mean concentrations, on the relations between concentration and flammability characteristics (ignition energy, time delay) and on the relevant frequencies/probabilities, including probabilities of ignition.

3. DISPERSION IN CLOSED PLANTS

The overall objective is to specify a method of estimating the hazardous area for dispersion in a closed plant, given a method of handling emission. The detailed objectives are to:

1. Specify a set of emission scenarios and associated frequency/ probability
2. Select methods of estimating vaporisation from pools in a closed space
3. Select methods of estimating dispersion in a closed space
4. Devise methods of defining hazardous area of gas cloud in a closed space
5. Devise methods of estimating effect of ventilation rates on (2–5)
6. Specify a standard approach to hazardous areas in a closed plant.

The work will follow a pattern broadly similar to that of Topic 2. It is believed, however, that there are fewer correlations developed specifically for vaporisation and dispersion indoors than to outdoors and it will therefore be necessary to take a view on whether or not the latter can be satisfactorily adapted.

The work will explore the possibility of the use of tracer tests as a complement to theoretical models for cases involving existing closed plants.

4. MINIMUM HAZARDOUS INVENTORY

The overall objective is to determine whether a minimum inventory can be specified below which the techniques of area classification is not appropriate and, if so, to specify this inventory. The detailed objectives are to:

1. Specify a set of inventory scenarios
2. Specify a minimum inventory (if appropriate).

The work will start with first principles but will move at an early stage to discussion with industry. It will seek to devise criteria for treating a hazard as negligible based on consequences and frequency/probability, and to specify a set of scenarios for different operating and storage conditions, to specify a minimum inventory for each scenario, provided the criteria are met, and to determine from these results whether, for simplicity, a single minimum inventory can be specified and whether a distinction can be made between process and storage.

5. ZONE DEFINITION

The overall objective is to propose a practical approach to zone definition. At present the two principal candidate approaches appear to be:

1. To use quantitative methods in each individual case (Method 1)

or

2. to use a set of rules which in use involves no, or little, quantification although they may be based on prior quantitative methods (Method 2).

The detailed objectives are to:

1. Define risk criteria
2. Utilise the results of Topics 1–3 to devise a quantitative method of zone definition (Method 1)
3. Utilise the results of Topics 1–3 to devise a non-quantitative method of zone definition (Method 2)
4. Compare (2) and (3) and recommend preferred method.

The work will start from the results of Topics 1–3 but will move as soon as practicable to discussions with the IIGCHL and with industry.

Where calculation methods are proposed, recommendations on the 'best buy(s)' will be given.

The work will be done bearing in mind that it will be more valuable if the information is presented in a form which can be used by those formulating codes of practice and those devising programs for computer aided design so that both draw from a common source.

PROJECT TIMETABLE

The proposed timetable for the practical elements of the project is as follows:

Stage 1 (year 1)
Literature surveys
Discussions with interested parties
Collection of data
Examination and selection of models
Stage 2 (year 2)
Development of methodologies
Development of overall approach
Stage 3 (year 3)
Further development of methodologies
Testing of and discussion of overall approach
Final proposals

REPORTING ARRANGEMENTS
Progress will be reported by progress reports at three monthly intervals and by meeting with the Steering Committee as appropriate.

It is envisaged there will probably be two final reports, one dealing with the project as a whole and the other covering guidance notes on the methodology.

A.2 CONDUCT OF PROJECT
As envisioned in the original terms of reference, areas of difficulty were notified to the Steering Committee as the project proceeded. It became clear that the main areas of difficulty in the development of a quantitative approach are those associated with the fundamental philosophy to be adopted and with the areas of leaks and ignition and effort has been concentrated on these, resulting in the case of the latter in the development of the fire and explosion model. It has not proved possible to progress as far as originally hoped with the development of an immediately applicable methodology or with the questions of zone distance, dispersion under ventilated conditions and of minimum inventory.

APPENDIX B—COMPOSITION OF GUIDING COMMITTEES

During the period of the project the membership of the Inter-Institutional Group on Classification of Hazardous Locations was

G.A.Lee (chairman)
B.J.Church   Institution of Gas Engineers
P.C.Palles-Clark   Institution of Electrical Engineers
S.J.Duke   Institution of Electrical Engineers
R.Dyson   Institution of Gas Engineers
B.W.Eddershaw   Institution of Mechanical Engineers
I.L.Edwards   Institute of Petroleum
K.W.Gladman   Institution of Mechanical Engineers
M.Pantony   Health and Safety Executive
K.Ramsey   Health and Safety Executive (to April, 1987)
C.A.W.Townsend   Institution of Chemical Engineers (from April, 1987)
H.B.Whitehouse   Institution of Chemical Engineers

The project has been guided by a Steering Committee the membership of which was

K.W.Ramsey (chairman) (to April, 1987)
C.A.W.Townsend (chairman) (from April, 1987)
P.C.Palles-Clark
R.Dyson
B.W.Eddershaw
K.W.Gladman
M.Pantony

M.L.Ang (to August 1987) Loughborough University of Technology
A.W.Cox   Loughborough University of Technology
F.P.Lees   Loughborough University of Technology

APPENDIX 1—RELEVANT STANDARDS AND CODES

Relevant standards and codes include the following:

BRITISH STANDARDS AND CODES

BS 5345: *Code of practice for the selection, installation and maintenance of electrical apparatus for use in potentially explosive atmospheres* (*other than mining applications or explosive processing and manufacture*). *Part 2: Classification of hazardous areas* (1983)

Institute of Petroleum Model Code of Safe Practice. Part 1: Electrical (1965)

Institute of Petroleum Model Code of Safe Practice. Part 8: Drilling and Production in Marine Areas (1964)

Electrical Installations in Flammable Atmospheres. ICI/RoSPA (1975)

Code of Practice for Hazardous Area Classification for Natural Gas. British Gas Engineering Standard BGC/PS/SHA1 (1986)

CS 2: *The Storage of Highly Flammable Liquids.* Health and Safety Executive (1978)

CS 4: *The Keeping of LPG in Cylinders or Similar Containers.* Health and Safety Executive (1981)

CS 5: *The Storage of LPG at Fixed Installations.* Health and Safety Executive (1981)

PM 25: *Vehicle Finishing Units: Fire and Explosion Hazards.* Health and Safety Executive (1981)

Bulk Storage and Handling of High Strength Potable Alcohol. Health and Safety Executive (1986)

Highly Flammable Liquids in the Paint Industry. Health and Safety Executive (1978)

Sludge Digestion Plant: Zoning for Hazardous Areas. Severn Trent Water Authority (1988)

OVERSEAS AND INTERNATIONAL STANDARDS AND CODES

IEC 79-10: *Classification of Hazardous Areas.* International Electrotechnical Commission (1979)

Code for the Construction and Equipment of Mobile Offshore Drilling Units (MODU Code). International Maritime Organisation (1980)

API 500A: *Classification of locations for electrical installations in petroleum refineries, 4th ed.*, American Petroleum Institute (1982)

API 500B: *Classification of areas for electrical installations at drilling rigs and production facilities on land and on marine fixed and mobile platforms, 2nd ed.*, American Petroleum Institute (1973)

API 500C: *Classification of areas for electrical installations at petroleum and gas pipeline transportation facilities, 2nd ed.*, American Petroleum Institute (1984)

NFPA 30: *Flammable and Combustible Liquids Code.* National Fire Protection Association (1984)

NFPA 70: *National Electrical Code.* National Fire Protection Association (1987)

Guidelines for the Prevention of Danger in Explosive Atmospheres with Examples. Berufsgenossenschaft der chemischen Industrie (Federal Republic of Germany) (1975)

Guidelines for the classification of hazardous areas in relation to gas explosion hazards and to the operation and selection of electrical apparatus. Directorate General of Labour (Netherlands) (1979)

Rules for the management and usage of LPG stores. Comité Professionel du Pétrole (France) (1972)

Rules for the management and usage of liquid hydrocarbons. Comité Professionel du Pétrole (France) (1972)

CEI 64–2: *Code for Electrical Installations in Locations with Fire and Explosive Hazards, 7th ed.*, Comitato Elettrotecnico Italiano (Italy) (1974)

Area Classification and Ventilation. Det Norske Veritas (Norway), Offshore Installations Technical Note TN B302 (1981)

Swedish Standard SS 421 08 20: *Classification of Hazardous Areas* (1984)

AS 2430.1: *Classification of Hazardous Areas. Part 1 Explosive Gas Atmospheres.* Standards Association of Australia (1987)

AS 2403.3: *Classification of Hazardous Areas. Part 3 Specific Occupancies.* Standards Association of Australia (1987)

AS 1482: *Electrical Equipment for Explosive Atmospheres – Protection by Ventilation - Type of Protection V.* Standards Association of Australia (1985)

Note:
(a) The Institute of Petroleum currently has in draft a code on hazardous area classification, which has been separated from the electrical code.

APPENDIX 2—FIRE AND EXPLOSION SURVEY OF NATIONAL CASE HISTORIES

Data on the distribution of the leak sources and the ignition sources involved in fire and explosion incidents are not readily available, but a limited amount of information has been obtained.

The principal source is a leak and ignition study made by the authors using statistics on national incidents provided by the Health and Safety Executive. The data are from an HSE data bank which is based on reports of notifiable injuries and dangerous occurrences. The study covers the one-year period April 1987 – March 1988.

A summary of the leak and ignition incidents is shown in Table A2.1. The 968 incidents shown in Table A2.1 cover a range of industrial situations from catering to process plants and the combustible materials include solids as well as flammable fluids. For the present study only those incidents have been retained which were judged to relate to process plant and to situations where hazardous area zoning appears applicable. There were some 345 such incidents.

However, of these some 68 incidents were hotwork classed under the hotwork entry and another 42 hotwork classed under other entries, making 110 hotwork incidents altogether.

A classification as hotwork is potentially ambiguous. It may mean that hotwork was the initiating event or that a leak occurred and was subsequently ignited by hotwork. It has been assumed that in all cases hotwork was the initiating event and the hotwork cases have therefore been discarded. The justification for this is as follows. Many of the incidents relate to items such as tanks or drums which suggest that the incident involved welding or cutting and ignition of flammable materials in the container. Support for this interpretation is given in the work of Forsth (1981a) on ignition incidents in the Norwegian North Sea. He gives some 80 hotwork incidents but all are cases where hotwork was the initiating event.

Table A2.1   Fire and explosion survey of national case histories: summary of incidents

	No. of incidents	No. of injuries Fatalities	Total injuries[a]
Flames: general[b]	237	2	224
LPG fired equipment	24	0	15
Hot surfaces	48	1	31
Friction	36	0	15
Electrical	70	1	54
Hot particles	20	0	17
Static electricity	19	0	10
Smoking	38	3	38
Autoignition	25	0	7
Other	5	0	5
Unknown	300	5	215
Spontaneous ignition, etc.[c]	26	0	14
Hotwork	120	4	101
Total	968	16	746

Notes:
(a) Including fatalities.
(b) Excluding LPG fired equipment.
(c) Including pyrophoric ignition, etc.

Further, there are some 10 incidents classed under the spontaneous ignition entry, which covers reaction runaways, pyrophoric ignitions, etc. Again, these are not relevant to the present study of external ignition sources and have been discarded.

There are therefore some 225 relevant incidents. These have been divided into two groups. The first group is those occurring due to leaks from essentially closed process plant, the second those occurring on plant and activities with open surfaces, and also mobile plant and transport. The numbers of incidents in these two categories are

Closed process plant	86
Plant and activities with open surfaces	139

For closed process plant Tables A2.2 and A2.3 give, respectively, summaries of incidents and injuries and Tables A2.4 and A2.5 the distribution of leak sources and ignition sources.

For plant and activities with open surfaces, etc., Tables A2.6 and A2.7 give, respectively, summaries of incidents and injuries and Tables A2.8 and A2.9 the distribution of leak sources and ignition sources.

Table A2.2   Fire and explosion survey of national case histories: incidents on closed process plant

	Plant	Reactor	Vessel	Tank	Heat exchanger	Vaporiser	Pump	Pipework	Hose	Total
					No. of incidents					
Flames:										
General	3	-	1	1	-	-	1	2	-	8
LPG fired equipment	-	-	-	-	-	1	-	1	-	2
Hot surfaces	1	1	1	1	1	-	1	3	1	10
Friction	3	-	-	-	-	-	1	-	-	4
Electrical	-	-	-	-	-	-	5	3	-	8
Hot particles	1	-	-	1	-	-	1	-	-	3
Static electricity	1	2	2	1	-	-	-	-	-	6
Smoking	-	-	-	-	-	-	-	-	-	0
Autoignition	1	1	1	-	2	-	1	1	-	7
Unknown	9	4	5	3	1	-	4	7	5	38
Total	19	8	10	7	4	2	13	17	6	86

Table A2.3   Fire and explosion survey of national case histories: injuries on closed process plant

	Plant	Reactor	Vessel	Tank	Heat exchanger	Vaporiser	Pump	Pipework	Hose	Total
					No. of injuries					
Flames:										
General	2(1)	-	1	1	-	-	-	-	-	4(1)
LPG fired equipment	-	-	-	-	-	-	-	-	-	0
Hot surfaces	-	-	-	1	-	-	1	-	-	2
Friction	-	-	-	-	-	-	-	-	-	0
Electrical	-	-	-	-	-	-	-	-	-	0
Hot particles	-	-	-	-	-	-	-	-	-	0
Static electricity	-	-	-	-	-	-	-	-	-	0
Smoking	-	-	-	-	-	-	-	-	-	0
Autoignition	-	-	-	-	-	-	-	-	-	0
Unknown	-	1	-	-	-	-	1	-	-	2
Total	2(1)	1	1	2	-	-	2	-	-	8(1)

Notes:
(a) Figures are total injuries, including fatalities. Fatalities are given in brackets

Table A2.4   Fire and explosion survey of national case histories: leak sources on closed process plant

	No. of incidents	Proportion of incidents (%)
Plant	19	22.1
Reactor	8	9.3
Vessel	10	11.6
Tank	7	8.1
Heat exchanger	4	4.7
Vaporiser	2	2.3
Pump	13	15.1
Pipework	17	19.8
Hose	6	7.0
Total	86	100.0

Table A2.5   Fire and explosion survey of national case histories: ignition sources on closed process plant

	No. of incidents	Proportion of incidents (%)
Flames: general	8	9.3
LPG fired equipment	2	2.3
Hot surfaces	10	11.6
Friction	4	4.7
Electrical	8	9.3
Hot particles	3	3.5
Static electricity	6	7.0
Smoking	-	-
Autoignition	7	8.1
Unknown	38	44.2
Total	86	100.0

Table A2.6   Fire and explosion survey of national case histories: incidents for plant and activities with open surfaces

	Solvent evaporating oven	Spray booth	Small container	Cleaning/ degreasing process	Tanker/ mobile plant	Other	Total
			No. of incidents				
Flames:							
General	2	1	9	4	4	7	27
LPG fired equipment	-	-	1	-	-	1	2
Hot surfaces	-	2	2	3	6	7	20
Friction	-	4	-	1	-	6	11
Electrical	2	2	4	1	11	9	29
Hot particles	-	-	-	-	-	-	0
Static electricity	-	2	3	1	-	4	10
Smoking	-	1	5	3	3	5	17
Autoignition	-	-	-	1	-	1	2
Unknown	-	3	7	1	9	1	21
Total	4	15	31	15	33	41	139

Table A2.7   Fire and explosion survey of national case histories: injuries for plant and activities with open surfaces

	Solvent evaporating oven	Spray booth	Small container	Cleaning/ degreasing process	Tanker/ mobile plant	Other	Total
			No. of injuries[a]				
Flames:							
General	1	-	6	2	3	1	13
LPG fired equipment	-	-	-	-	-	1	1
Hot surfaces	-	1	-	-	5(1)	1	7(1)
Friction	-	1	-	-	-	1	2
Electrical	-	-	-	-	4(1)	-	4(1)
Hot particles	-	-	-	-	-	-	0
Static electricity	-	-	-	-	-	-	0
Smoking	-	-	4	2	1	1	8
Autoignition	-	-	-	-	-	-	0
Unknown	-	-	2	-	4	-	6
Total	1	2	12	4	17(2)	5	41(2)

Notes:
(a) Figures are total injuries, including fatalities. Fatalities are given in brackets.

Table A2.8   Fire and explosion survey of national case histories: leak sources for plant and activities with open surfaces

	No. of incidents	Proportion of incidents (%)
Solvent evaporating oven	4	2.9
Spray booth	15	10.8
Small container	31	22.3
Cleaning/degreasing process	15	10.8
Tanker/mobile plant	33	23.7
Other	41	29.5
Total	139	100.0

Table A2.9   Fire and explosion survey of national case histories: ignition sources for plant and activities with open surfaces

	No. of incidents	Proportion of incidents (%)
Flames: general	27	19.4
LPG fired equipment	2	1.4
Hot surfaces	20	14.4
Friction	11	7.9
Electrical	29	21.0
Hot particles	-	-
Static electricity	10	7.2
Smoking	17	12.2
Autoignition	2	1.4
Unknown	21	15.1
Total	139	100.0

APPENDIX 3—INCIDENT AND INJURY RATES

A3.1 EMPLOYMENT AND ACCIDENT STATISTICS

The Health and Safety Statistics 1985–86 give statistics on fatal and major injuries in manufacturing industries for the period 1981–85. The definition of major injury used for this period is that given in the Notification of Accidents and Dangerous Occurrences Regulations (NADOR) 1980. Some principal statistics are shown in Table A3.1, Sections A and B.

Table A3.1   Some statistics on accidents in manufacturing industries and in the chemical and oil industries

A Manufacturing industries, 1981–1985: fatal and major injuries

Industry	Fatal and major injuries											
	1981		1982		1983		1984		1985		Total	
	F	M	F	M	F	M	F	M	F	M	F	M
Chemical industry	8	321	6	344	10	374	5	370	5	390	34	1799
Mineral oil processing	3	28	3	36	1	29	1	24	1	24	9	141
Total	11	349	9	380	11	403	6	394	6	414	43	1940

B Manufacturing industries, 1981–1985: fatal and major injury rates

	Fatal and major injury rates (incidence per 10^5 employees)				
	1981	1982	1983	1984	1985
Chemical industry	89.4	100.3	115.2	112.7	117.2
Mineral oil processing	108.4	154.2	136.4	130.2	139.7

C Chemical and oil industries, 1984

Industry	No. employed	Fatalities	Major injuries	Fatal and major injuries per 10^5 persons
Chemical industry	360,000	5	349	98.4
Other chemical processes	38,200	1	38	102.1
Mineral oil refining	18,200	1	14	82.4

Sources: Health and Safety Executive (1986); HM Chief Inspector of Factories (1986)

The report also records that in manufacturing industries in 1985 there were 89 major injuries due to fire and explosion. The total number of major injuries was 4743 so that fire and explosion injuries constituted 1.9%.

The 1984 Annual Report of HM Chief Inspector of Factories contains an article describing the work of the Chemical Industry National Group of the HSE. Some principal statistics for the chemical and oil industries are shown in Table A3.1, Section C.

This report also gives a breakdown on the numbers employed in various sections of the chemical industry. For 1985 the numbers employed were:

Pharmaceuticals, toilet preparation, soaps and detergents 111, 000
Other chemical industry 249, 000
Other chemical processes 38, 000
Mineral oil processing 18, 000

A3.2 NUMBER OF PERSONS EXPOSED

An estimate of the number of persons to fire and explosion hazard in the chemical and oil industries has been made as follows.

It is assumed that for the mineral oil industry the proportion of employees exposed is about one third and that for the chemical industry it is about one quarter. Then the number of persons exposed is

$$\text{No. exposed in the chemical industry} = (360, 000 + 38, 000 - 111, 000)/4$$
$$= 71, 750$$

$$\text{No. exposed in oil industry} = 18, 000/3 = 6000$$

$$\text{Total no. exposed} = 71, 750 + 6000 = 77, 750$$
$$= 80, 000, \text{ say}$$

A3.3 APPLICABLE FATALITIES

From data made available by the Health and Safety Executive there were in the period 1981–87 in the chemical and oil industries 6 fatalities attributable to ignition of flammable fluids. There were only five incidents, since one involved two deaths.

Three of the incidents, including the double fatality, occurred during maintenance work in refineries. In one case the ignition source was attributed to burners, in another to an air compressor and in the third it was not identified. The other incidents were during switch loading of a tanker, where ignition was attributed to static electricity, and an indoor release due to failure of cooling water on a condenser, for which the ignition source is not given.

The fire and explosion survey of national case histories for the one-year period 1987–88 described in Appendix 2 gives one fatality on closed process plant due to ignition of a flammable leak.

The total number of fatalities from all causes as given in Table A3.1 are 11, 9, 11, 6, 6 for the years 1981 to 1985, respectively. The proportion of major injuries attributable to fire and explosion was 1.9%.

On the basis of the first set of data, those for 1981–87, there were in the chemical and oil industries 6 deaths over a period of 7 years, giving an average of 0.86 per year.

A3.4   FATAL ACCIDENT RATE

The Fatal Accident Rate (FAR) is defined as the number of fatalities per 10^8 exposed hours.

It is assumed that an exposed employee works a 40 hour week for 48 weeks per year. Then

No. of exposed hours per employee per year $= 40 \times 48 = 1920$

If the total number of exposed employees in the chemical and oil industries is 80, 000, then

No. of exposed hours of all employees per year $= 1920 \times 80, 000$
$$= 1.54 \times 10^8$$

Then using the tentative estimate of 0.86 fatal accidents per year due to ignition of a flammable leak

Fatal Accident Rate $= 0.86/(1.54 \times 10^8) = 0.56$ fatalities/10^8 exposed hours

This figure for the FAR should be regarded as provisional.

A3.5   INCIDENT FATALITY AND INJURY RATE

An estimate of the probability of injury or fatality per incident may be obtained from the fire and explosion survey of national case histories. For incidents involving ignition of a flammable leak from closed process plant there were 86 incidents, one fatality and 7 other major injuries. This gives

Probability of fatality $= 1/86 = 0.012$ fatalities/incident

Probability of major injury $= 8/86 = 0.093$ major injuries/incident

Since over two-thirds of the incidents reported in the survey as a whole involved some form of injury, mostly minor injury, the number of incidents is not synonymous with the number of ignited leaks.

Figures for the incident fatality and injury rates have been given by Forsth (1981a, b) for the Gulf of Mexico (GoM) and the Norwegian North Sea (NNS). These are:

	GoM	NNS
Fatalities/incident	0.24	0.04
Injuries/incident	0.96	0.04

Neither set of figures is satisfactory. The large discrepancy between the GoM and NNS figures is a strong indicator of underreporting for the GoM, while for the NNS the fatality figure is based on 5 fatalities which all occured in the same incident; the injury figure is more acceptable, being based on 5 incidents with one injury in each.

The NNS figures, however, are broadly comparable with those for the fire and explosion survey of national case histories.

The fatality and major injury rates per ignited leak are therefore almost certainly appreciably less than the rates per reported incident just quoted. Surveys of movements of people on plant have been done in support of hazard analysis. It is a common assumption in such studies that the probability of a man being near enough to be injured if an explosion occurs on an item of plant is 0.1.

A further estimate of the probability of fatality or injury might be made using the national fatality and injury data together with estimates such as those made in Appendix 14 of the number of ignited leaks per plant per year and with estimates such as those given in Appendix 6 of the number of plants.

APPENDIX 4—SOME IGNITION CASE HISTORIES

Some ignition incidents reported to the HSE in the period 1981–85 include the following:

1. The cooling water on a condenser on a distillation plant inside a flameproof room failed and large quantities of vapour escaped. Workers were under instructions to evacuate on cooling water failure, but the deceased person had gone back into the building. The vapour found a source of ignition and an explosion occurred. Deaths: 1.

2. A leak of LPG occurred on a tank when a tap was accidentally knocked. The tap was turned off 3–4 min later. Workers evacuated but returned after 20 min. An explosion then occurred which demolished the whole building. Deaths: 2.

3. Fire and explosion occurred on a road tanker during a filling operation. Diesel fuel was being loaded after a previous load of petrol, ie switch loading. There were no precautions for earthing or for limitation of filling flow rate. The ignition source may well have been static electricity. The deceased person was the tanker driver who was on top of the tanker carrying out the filling operation. Deaths: 1.

4. LPG leaked from an aerosol filling machine in a ventilated booth inside a ventilated filling room. The operator noticed the leak and went for the fitter, but did not turn the LPG off. When the fitter opened the filling room door, vapour 'flooded out' to a nearby shrink wrapping machine and ignited. There was fire described as a fireball. Deaths: 1.

Of these four cases three involved indoor plants and one a road tanker.

APPENDIX 5—INVENTORY OF LEAK SOURCES

A5.1 LITERATURE DATA

Some information on the inventory of leak sources is available in the literature. The following sources give some data on the number of components, length of pipework, etc, in various types of plant.

Hooper (1982) in a paper to assist cost estimation gives a data base showing the quantity of pipes of different diameters and number of fittings, flanges, valves in a representative selection of plants. The data base described is drawn from a representative selection of plants of different sizes in Monsanto. The total length of pipework of each diameter is given together with the associated number of fittings (elbows, tees, reducers, flanges, unions and couplings, caps and plugs) and valves. A selection of his data is given in Table A5.1.

Wetherold *et al.* (1983) in a paper on fugitive emissions give data on the number of valves in a refinery, an olefins plant and a cumene process unit. These data are given in Table A5.2.

Lipton and Lynch (1987) in a book covering fugitive emissions give data on the number of flanges, valves, pump and compressor seals, relief valves and drains in a large refinery. The data are based on work by the Environmental Protection Authority (EPA). These data are given in Table A5.3.

Wallace (1979) in a paper on fugitive emissions gives data on the number of flanges, valves, pump seals, compressor seals and relief valves in a medium sized plant. These data are given in Table A5.4.

Hughes, Tierney and Khan (1979) in a paper on fugitive emissions give data on the number of flanges, valves, pumps in 4 chemical plants. These data are given in Table A5.5.

Table A5.1   Inventory of leak sources: a data base of pipework fittings and valves (after Hooper, 1982)

Nominal pipe diameter (in)	Total length of pipe (ft)	Number of fittings	
		Flanges	Valves
1/2	33,990	1,818	11,589
3/4	33,123	2,973	7,551
1	124,513	12,552	10,363
1 1/2	121,212	7,299	3,313
2	142,891	11,727	4,199
3	125,550	10,427	2,441
4	84,705	6,608	1,346
6	77,717	4,578	898
8	67,667	3,592	466
10	39,225	1,613	301
12	16,445	762	162
14	3,997	342	72
16	10,292	506	90
18	3,530	·362	41
20	5,698	804	34
24	5,983	357	40
30	3,121	255	13
36	1,608	66	12

Source: Hooper (1982), Table on p.128.

Table A5.2   Inventory of leak sources: number of valves in a refinery and two other plants (after Wetherold, 1983)

A No. of valves in three plants

	No. of valves
Large integrated refinery	21,800
Large olefins plant	15,000
Cumene process unit	1,179

B No. of valves on different duties in large refinery

Valves	No. of items
Gas and light liquid only	13,334
All	21,776

Source: Wetherold (1983), Tables 1 and 3.

Table A5.3   Inventory of leak sources: estimated number of leak sources in a large refinery (after Lipton and Lynch, 1987)

Leak source	No. of items
Flanges	46,500
Valves	11,500
Pump seals	350
Compressors	70
Relief valves	100
Drains	650

Source: Lipton and Lynch (1987), Table 7.2

Table A5.4   Inventory of leak sources: estimated of number of leak sources on a medium sized plant (after Wallace, 1979)

Leak source	No. of items
Flanges	2,410
Valves	
In-line, gas	365
In-line, liquid	670
Open-ended	415
Pump seals	
Packed	6
Mechanical, single	43
Mechanical, double	10
Compressor seals	2
Safety relief valves	50

Source: Wallace (1979), p.92.

Table A5.5   Inventory of leak sources: number of leak sources on four petrochemical plants (after Hughes, Tierney and Khan, 1979)

Leak source	No. of items Monochloro-benzene plant	Butadiene plant	Ethylene oxide/glycol plant	Dimethyl terephthalate plant
Flanges	1,500	26,000	NA	NA
Valves	640	6,700	NA	NA
Pumps	25	174	69	67

Source: Hughes, Tierney and Khan (1979), Table 3.

Although restricted to valves inspected rather than total number of valves, the information given by Morgester *et al.* (1979) is useful in obtaining an impression of the distribution of valves across the units of a refinery. Table A5.6 gives the number of valves inspected by these authors in 6 refineries.

Table A5.6 Inventory of leak sources: number of valves inspected in 6 refineries (after Morgester *et al.* 1979)

Unit	No. of valves inspected
Storage	2,398
Reformer	1,803
FCC	1,737
Fractionation	1,540
Alkylation	1,522
Crude unit	1,243
Coker	870
LSFO	726
Hydrotreating	637
Hydrocracker	609
Other units	60
Total	13,685

The HSE (1978) in the first Canvey Report give data on the length of LPG pipework in a refinery (see Lees (1980), p.1021).

A5.2 MATERIAL RUNOFFS

It was considered desirable to supplement this information by further studies of inventories of plant equipment and pipework, partly to obtain a more comprehensive coverage of different types of plant and partly to obtain data on leak sources not covered.

Information has been made available to the project on the following types of plant:

Petrochemical plant

Oil production platform

Inventory data for a petrochemical plant have been made available by Company A in the form of a handover document of material takeoff for Plant A, a medium sized chemical plant.

Inventory data for an oil production platform have been made available by Company B in the form of a computer printout of material takeoff for plant B, an oil production platform.

Table A5.7 gives a summary of data for pipework, flanges and valves and pumps on Plant A and Tables A5.8 and A5.9 a summary of similar data for Plant B.

Table A5.7 Inventory of leak sources: pipework, flanges and valves in a medium sized chemical plant (Plant A)

Pipe diameter (mm)	Length of pipe (m)	(%)	Gaskets	Flanges	Valves (total)	Flanged joints	Valves (process)
15	290	2.0	132	173	537	102	177
20	108	0.75	250	394	203	192	102
25	2404	16.7	821	911	787	632	590
40	2046	14.2	309	367	112	238	112
50	1860	12.9	701	703	225	539	225
80	4434	30.7	748	538	180	575	180
100	762	5.3	239	252	64	184	64
150	750	5.2	187	181	33	144	33
200	552	3.8	132	127	15	102	15
250	322	2.2	67	72	9	52	9
300	66	0.46	38	34	5	29	5
350	24	0.17	6	8	-	5	-
400	672	4.7	25	21	-	19	-
450	30	0.21	16	18	2	12	2
500	90	0.62	8	8	-	6	-
600	12	0.03	6	6	-	5	-
Total	14422		3685	3813	2172	2836	1514

Table A5.8 Inventory of leak sources: pipework, flanges and valves on an oil production platform (Plant B)

Pipe diameter (in)	(mm)	Length of pipe (m)	(%)	Gaskets	Flanges	Valves (total)	Flanged joints	Valves (process)
0.5	15	2292	4.9	1517	1737	394	1011	130
0.75	20	253	0.54	1689	1056	597	1126	298
1	25	10317	22.1	2948	2564	886	1965	664
1.5	40	6956	14.9	1638	1462	411	1092	411
2	50	7334	15.7	3014	2037	661	2009	611
3	80	5224	11.2	1478	1431	221	985	221
4	100	4256	9.1	1491	1595	203	994	203
5	125	576	1.2	163	152	7	109	7
6	150	3830	8.2	1341	1266	265	894	265
8	200	1990	4.3	766	725	111	511	111
10	250	1108	2.4	463	409	67	309	67
12	300	765	1.6	193	109	17	129	17
14	350	194	0.42	70	60	6	47	6
16	400	824	1.8	212	212	9	141	9
18	450	210	0.45	78	80	-	52	-
20	500	457	1.0	72	83	7	48	7
Total		46586		17133	14978	3862	11422	3027

Table A5.9   Inventory of leak sources: distribution of pipework by fluid carried on an oil production platform (Plant B)

Pipe diameter (mm)	Proportion on gas or gas and liquid (%)
15	0
20	0
25	0.5
40	5
50	7
80	10
100	9
150	7
200	12
250	38
300	8
350	16
400	21
450	21
500	13

The data given in these tables have been used in Section 8 to estimate the length of pipework and the number of flanges, valves and pumps on a 'standard' plant.

It is also necessary to know the number of small bore connections. For a standard plant an estimate of 450 has been made, based principally on data on instruments given by Anyakora, Engel and Lees (1971) which covered plants estimated as equivalent to some 12 standard plants.

APPENDIX 6—INVENTORY OF MAJOR PLANTS

In order to estimate the frequency of an event such as a fire or vapour cloud explosion it is necessary to make an estimate of the number of installations at risk.

The installations concerned have been taken as chemical and petrochemical plants, refinery units, LPG installations and natural gas installations, which together are referrred to as major installations handling flammables. An estimate of the number of such installations has been made for the US, for Western Europe and for the UK.

For chemical and petrochemical plants information on the number of plants in the US and in Western Europe (WE) and on their individual capacities is given in the following publications by SRI International:

Directory of Chemical Producers – United States (1988)
Directory of Chemical Producers – Western Europe (1988)

For refineries information on the number of refineries is given in the International Petroleum Encyclopaedia. Further useful information is given in periodic surveys of refineries in the *Oil and Gas Journal.*

Figures for the number of installations notifiable under the Notification of Installations Handling Hazardous Substances (NIHHS) Regulations 1982 have been given by the HSE (Pape and Nussey, 1985; Pape 1989). These include data on the number of LPG installations and also provide a crosscheck on the other types of installation.

For chemical and petrochemical plants from the SRI data

No. of chemical/petrochemical plants in US = 1420
No. of chemical/petrochemical plants in WE = 1455
No. of chemical/petrochemical plants in UK = 185

Also for 20 representative products
Ratio of number of plants:
US/WE = 0.92
UK/WE = 0.12

Ratio of plant capacities
US/WE = 1.08
UK/WE = 0.12

For refineries from the International Petroleum Encyclopaedia for 1983 and from a survey of refineries for 1986 (Anon., 1985)
No. of refineries in non-Communist world, excluding US = 428
No. of refineries in US = 225
No. of refineries in WE = 120
No. of refineries in UK = 15

It is estimated that the number of major units in a refinery is on average about 5.

Hence for chemical/petrochemical plants and refinery units
No. of plants in US = 1420 + 5 × 225 = 2545
No. of plants in WE = 1455 + 5 × 120 = 2055
No. of plants in UK = 185 + 5 × 15 = 260

According to Pape and Nussey (1985) and Pape (1989) there are some 1600 sites notifiable under the NIHHS Regulations, of which there are
LPG installations = 600
Natural gas installations = 400
Chlorine installations = 120

Of the LPG installations some 130 are cylinder storages and 450 LPG storage installations with capacities greater than 25 te. Then by difference there are some 480 other installations. Some of these will be installations handling toxics such as ammonia or solid materials such as ammonium nitrate.

Then for the UK the number of major installations handling flammables is estimated as
Chemical/petrochemical plants and refinery units = 260
LPG installations = 450
Natural gas installations = 400
Total installations = 260 + 450 + 400 = 1110
Ratio of total installations to chemical/petrochemical plants and refinery units = 1110/260 = 4.27

Applying the same ratio to the figures for the US and for WE
No. of major installations handling flammables in US = 4.27 × 2545
= 10867

No. of major installations handling flammables in WE $= 4.27 \times 2055$
$$= 8775$$

No. of major installations handling flammables in
US and WE $= 10867 + 8775 = 19642$

The estimates just given involve a number of assumptions, such as the constancy of ratios of certain types of installation, which will not hold exactly, but it is considered that the estimates obtained are sufficiently accurate for present purposes.

However, not all the major installations handling flammables appear to present the same risk of fires and explosions. The numbers given above are inflated by the inclusion of large numbers of LPG and natural gas installations, which figure very little in the fire and explosion statistics. For example, there are in the vapour cloud explosion incidents described by Davenport (1977, 1983) and considered in Appendix 14 very few in LPG or natural gas installations. Moreover, it is judged that some of the chemical and petrochemical plants have a low risk of such explosions. The best estimate of the number of plants in the UK which are broadly equivalent in fire and explosion risk to a refinery unit is some 180. Then

No. of refinery unit type plants in UK $= 180$

Ratio of refinery type plants to
major plants handling flammables in UK $= 180/1110 = 0.162$

No. of refinery unit type plants in US $= 0.162 \times 10867 = 1760$

No. of refinery unit type plants in WE $= 0.162 \times 8775 = 1422$

APPENDIX 7—STANDARD HOLE SIZES

In Section 10 the size of holes for leaks is discussed and standard hole classes are defined. This appendix gives some of the supporting background information relevant to determination of hole size.

A7.1 LEAK SOURCE DEFINITIONS

Data on the leak sources assumed in industry for the purpose of doing hazardous area classification may be given in various forms.

The most straightforward form is the actual hole area. In this case it is necessary to apply some coefficient of discharge. In some cases the area is quoted already incorporating the coefficient of discharge. This is referred to here as the modified area.

For seals on shafts the area may be determined if an annular hole, or clearance, is assumed. Often this clearance is itself taken to be proportional to the shaft diameter, so that in effect the hole size becomes proportional to the square of the diameter. Again it is not always clear whether a coefficient of discharge is already incorporated.

In some cases the leak is quoted on the quite different basis of the mass flow rate of the fluid. The pressure is not usually stated.

A7.2 SOME PARTICULAR LEAK SOURCES

FLANGE GASKET LEAKS

For a leak on a gasket the principal severe leak considered is usually the loss of a section of the gasket between two bolt holes. Information is therefore required on flanges and associated bolts for different pipe sizes. Two principal sets of piping standards are ANSI B16.5 and MSS-SP44 for pipes up to 24 in NB and over 24 in NB, respectively. Selected specifications from these standards are given in Table A7.1.

Table A7.1   Specifications of flanges and bolts for selected pipe sizes according to ANSI 16.5 and MSS-SP44

Class	Flange NB (in)	No. of holes	Hole diameter (in)	Bolt diameter (in)
ANSI 150	0.5–1.5	4	0.62	1/2
	2–3	4	0.75	5/8
	3.5–4	8	0.75	5/8
	5–8	8	0.88	3/4
	10–12	12	1.0	7/8
	14	12	1.12	1
	16	16	1.12	1
	18	16	1.25	9/8
	20	20	1.25	9/8
	24	20	1.38	5/4
ANSI 300	0.5	4	0.62	1/2
	0.75–1.25	4	0.75	5/8
	1.5	4	0.88	3/4
	2	8	0.75	5/8
	2.5–5	8	0.88	3/4
	6	12	0.88	3/4
	8	12	1.0	7/8
	10	16	1.12	1
	12	16	1.25	9/8
	14	20	1.25	9/8
	16	20	1.38	5/4
	18–20	24	1.38	5/4
	24	24	1.62	3/2
MSS 150	26	24	1.38	
	28	28	1.38	
	30	28	1.38	
	32	28	1.62	
	34	32	1.62	
	36	32	1.62	
MSS 300	26	28	1.75	
	28	28	1.75	
	30	28	1.88	
	32	28	2.0	
	34	28	2.0	
	36	32	2.12	

PUMP AND COMPRESSOR SEAL LEAKS

For pumps and compressors the leak size is usually quoted in terms of the shaft diameter. A typical relation is

$$A = \pi l d \qquad (A7.1)$$

where A is the leak area (mm^2), d the shaft diameter (mm) and l the clearance (mm). This therefore gives the actual hole area in a straightforward way.

In other cases the hole area is quoted as

$$A = k\pi d \qquad \text{(A7.2)}$$

where k is a constant. It is usually not specified whether the constant k is simply equivalent to the clearance l or whether it also incorporates a coefficient of discharge.

Another formula, at first sight rather puzzling, is

$$A = k'\pi d^2 \qquad \text{(A7.3)}$$

where k' is a constant. This type of equation occurs where a shaft diameter of a particular size is taken as a base case and incorporated into the constant k'. Thus

$$A = k\pi d \qquad \text{(A7.4)}$$

At the standard diameter d_s, $d = d_s$ and

$$A = k\pi d \cdot \frac{d}{d_s} \qquad \text{(A7.5a)}$$

$$= k'\pi d^2 \qquad \text{(A7.5b)}$$

where

$$k' = \frac{k}{d_s} \qquad \text{(A7.6)}$$

DRAIN AND SAMPLE POINT LEAKS
A typical sample point diameter is 20 mm. Drain sizes are rather wider in range. Typical diameters are 15, 25, 40 and 50 mm.

A7.3 STANDARD HOLE SIZE CLASSES
The standard hole size classes used here are as follows.

FLANGES

The principal types of flange are those with compressed asbestos fibre (CAF) and spiral wound joint (SWJ) gaskets and metal-to-metal ring-type joints (RTJ).

A typical thickness of a CAF gasket is 1.6 mm. Other thicknesses include 0.6 and 3 mm (Page and Nussey, 1985). The aperture given by a leak from a SWJ or RTJ flange is much less, a typical effective thickness being about 0.05 mm.

The hole size in a gasket failure may be that due to the complete section between bolt holes or something much smaller. In a metal-to-metal joint the hole may be due to scoring or pitting and is unlikely to extend over a whole section.

In industrial practice the hole size for a complete section failure of a gasket is usually calculated using the actual gasket thickness and sector between bolt holes.

For lesser holes a typical value for a CAF gasket is 2.5 mm^2 and for an RTJ 0.25 mm^2 or less, with the value for an SWJ gasket intermediate between these.

The approach to flange hole sizes adopted here is as follows. For a severe failure of a CAF gasket the hole width is taken as 1 mm and the hole size is determined using the sector between the bolts. For example, if the arc of the latter is 50 mm, the hole size is 50 mm^2. A smaller leak on a CAF gasket is taken as 2.5 mm^2.

For an SWJ gasket the hole width is taken as 0.05 mm. Then for sector failure with an arc of 50 mm the hole size is 2.5 mm^2. A smaller leak is taken as 0.25 mm^2.

For an RTJ the hole width is taken as 0.05 mm and the arc as 10 mm, giving a hole size of 0.5 mm^2. A smaller leak is taken as 0.1 mm^2.

VALVES

The hole sizes used in industrial practice are typically 0.25 mm^2 but with 2.5 mm^2 for more severe cases.

The hole size values adopted here are 0.25 mm^2 for normal duty valves and 2.5 mm^2 for severe duty valves and for large valves (> 150 mm diameter lines).

CENTRIFUGAL PUMPS

Centrifugal pumps with packed glands are not well suited to handling flammable liquids and only those with mechanical seals are considered here.

In industrial practice the hole size for a leak from a mechanical seal is typically determined from equation (A7.1) using an annular hole width associated with a particular shaft diameter and then assuming that the hole width is proportional to diameter as in equation (A7.2).

If a throttle bush is used, the hole size is reduced by a factor. The credit for this varies, the hole size reduction typically lying in the range 3–5.

The approach to pump hole size adopted here is as follows. A 25 mm shaft pump is taken as the base case. For such a pump the hole sizes adopted are

	Hole size (mm^2)
Mechanical seal, no throttle bush	25
Mechanical seal, throttle bush	5

For other pump sizes the hole size is taken as proportional to the square of the diameter.

RECIPROCATING COMPRESSORS

In industrial practice typical values of hole sizes for the various release points on reciprocating compressors tend to lie in the range 1–5 mm^2.

Here a hole size of 2.5 mm^2 is adopted for all leaks on reciprocating compressors.

CENTRIFUGAL COMPRESSORS

In industrial practice the hole size for a leak from a compressor seal is generally determined in a manner similar to that for pump seals. An annular hole size associated with a particular shaft diameter is used in equation (A7.1) and it is then assumed that the hole width is proportional to diameter as in equation (A7.2).

Seals may be purged labyrinth seals or floating ring seals. The latter tend to give a smaller hole size, the reduction factor typically being about 6.

The approach to centrifugal compressor hole size adopted here is as follows. A 150 mm diameter shaft compressor is taken as the base case. For this case the hole sizes adopted are:

	Hole size (mm^2)
Purged labyrinth seal	250
Floating ring seal	50

For other shaft sizes the hole size is taken as proportional to the square of the diameter.

SMALL BORE CONNECTIONS

For small bore connections the hole sizes used in industrial practice tend to lie in the range 0.1–1 mm^2.

The hole size for small bore connections adopted here is 0.25 mm^2.

DRAIN AND SAMPLE POINTS

For a drain or sample point use of the actual hole size is preferred. The corresponding standard hole size will generally be 100 or 250 mm^2.

APPENDIX 8—EQUIPMENT LEAK FREQUENCY

In this appendix a review is given of information available on the frequency of failures and of leaks on equipment. Some limited information is also given on leak size distributions.

A8.1 PIPEWORK FAILURE AND LEAK FREQUENCY

The following sources give some information on pipework failure frequencies, leak frequencies and leak size distribution.

Pipework failure frequencies may be given per unit length, per section or per plant.

Bush (1977)
Pipework failure frequencies per reactor, Table 2
Proportion of failures involving leaks, Table 3

First Canvey Report (HSE, 1978)
LPG pipework failure frequency per unit length – see Lees (1980), p.1021
LPG pipework failure frequency per refinery – see Lees (1980), p.1021 and Table A10.3

Wallace (1979)
Pipework, or rather flange, leak frequency per plant, Table VIII

Welker and Schorr (1979)
LNG plant pipework failure frequencies per unit length, Fig.7

Batstone and Tomi (1980)
Pipework failure frequencies per connection (by diameter)

Hansen, de Heer and Kortland (1980)
Pipework and equipment leak frequencies expressed as flange equivalent, Table 1

Lees (1980)
Pipework failure literature, Table A9.1
Pipework failure frequencies per plant year, Table A9.5
Leaks from flanges, gaskets, welds, Table A9.3

Arulanantham and Lees (1981)
Failure frequencies of pipework per plant, Table 8

Rijnmond (1982)
Pipework leak frequencies per pipe section (by diameter), Table IX.1, p.374

Hawksley (1984)
Pipework failure frequencies per unit length (by diameter), Fig.7

Pape and Nussey (1985)
Pipework failure frequencies per unit length for 1 in pipe in chlorine installation, Table 1 and p.379

The interpretation of the data from these various sources is quite difficult, since they are given in different forms and there is considerable variation between data expressed in a given form.

The information given in the Rijnmond Report on failure frequencies of pipework is shown in Table A8.1, Section A. Failure

Table A8.1   Failure frequency of pipework: Rijnmond Report data (Rijnmond Public Authority, 1982)

A Original data

Pipe diameter	Leak frequency (leaks/m h)	
	Catastrophic leaks	Significant leaks
(mm)		
< 50	10^{-10}	10^{-9}
> 50 < 150	3×10^{-11}	6×10^{-10}
> 150	10^{-11}	3×10^{-10}

B Derived values

Pipe diameter	Leak frequency (leaks/m y)	
	Catastrophic leaks	Significant leaks
(mm)		
< 50	10^{-6}	10^{-5}
> 50 < 150	3×10^{-7}	6×10^{-6}
> 150	10^{-7}	3×10^{-6}

frequencies derived from those in Section A but put on an annual basis are shown in Section B.

Pape and Nussey (1985) give the following failure rates for 25 mm pipe on chlorine installations:

Frequency of guillotine fracture $= 3 \times 10^{-7}$/m y

Frequency of pipe split $= 3 \times 10^{-6}$/m y

Table A8.2 shows a comparison of the frequency of a guillotine fracture for pipe diameters < 75 mm derived by some of these sources.

Table A8.2   Failure frequency of pipework: comparison of values for guillotine fracture

	Failure frequency (failures/m y)
Gulf[a]	3×10^{-7}
Cremer and Warner[a]	1×10^{-6}
Rijnmond Report	3×10^{-7}
Pape and Nussey (1985)	3×10^{-7}
Batstone and Tomi (1980)[b]	3×10^{-6}

Notes:
(a) Quoted by Hawksley (1984).
(b) Converted on assumption that a 'connection' is 10 m.

It is also necessary to estimate the distribution of leak sizes. A simple distribution which is sometimes used in risk assessments is that shown in Table A8.3, Section A.

Table A8.3   Pipework leak size distributions

A Simple distribution

Leak size	Proportion (%)
A	1
0.1A	9
0.01A	90

B Distribution given in Rijnmond Report industrial comment

Leak size	Leak description	Proportion (%)
A	Rupture	5
0.2A	Catastrophic leak	15
0.05A	Big leak	25
0.01A	Small leak	55

137

Another distribution shown in Section B of the table is that quoted in the industrial comment to the Rijnmond Report, attributed to the Gulf data.

On the basis of the foregoing, the pipework leak frequencies adopted for use in conjunction with the equivalent standard plant, described in Section 8, are as shown in Table A8.4.

Table A8.4 Failure frequency of pipework: values adopted for equivalent standard plant

Pipe diameter (mm)	Leak frequency (leaks/m y)		
	Rupture leaks	Major leaks	Minor leaks
25	10^{-6}	10^{-5}	10^{-4}
50	10^{-6}	10^{-5}	10^{-4}
100	3×10^{-7}	6×10^{-6}	3×10^{-5}
300	10^{-7}	3×10^{-6}	10^{-5}

A8.2 FLANGE FAILURE AND LEAK FREQUENCY

Information on flange failure is given by Smith (1985), the Rasmussen Report and Pape and Nussey (1985).

Pape and Nussey (1985) give the following failure rates for 25 mm pipe on chlorine installations:

Frequency of gasket failure: gasket 0.6 mm thick $= 3 \times 10^{-6}$/y
Frequency of gasket failure: gasket 3 mm thick $= 5 \times 10^{-6}$/y
These failure rates are for the loss of a section between adjacent bolts.

Comparison of the frequencies of gasket failure gives:

	Failure frequency (failures/y)
Pape and Nussey	4×10^{-5}
Smith: lower limit	0.00018
upper limit	0.0044
Rasmussen Report	0.026

On the basis of the foregoing, the flange leak frequencies adopted are as follows:

Frequency of gasket failure: section leak $= 10^{-4}$ leaks/y
Frequency of gasket failure: minor leak $= 10^{-3}$ leaks/y
The area A of a section leak is taken to be the cross-sectional area of the

part of the circumference between adjacent bolts and of the thickness of the gasket. The area of a minor leak is taken as 0.1A.

A8.3 VALVE FAILURE AND LEAK FREQUENCY

Information on valve failure is given by the following sources:

Anyakora, Engel and Lees (1971)
Valve failure frequency – see Lees (1980), Table 13.5
Proportion of failures involving leaks – see Lees (1980), Table 13.9

Rasmussen Report (1974)
Valve failure frequency – see Lees (1980), Table A9.3

Wallace (1979)
Valve leak frequency for fugitive emissions, Table VIII

UKAEA (various reports)
Valve failure frequency – see Lees (1980), Table 13.5

Rijnmond Report (1982)
Valve leak frequencies, Table IX.1, p.374
(control valves, motor operated valves, hand valves)

Summers-Smith (1982)
Valve leak rates, p.7/4

Aupied, Le Coguiec and Procaccia (1983).
Valve failure frequencies, Table 2
Proportion of failures involving leaks, p.139–140

The data given in the above sources include the following. Wallace (1979) gives for process valves for fugitive emissions

Leak frequency = 0.06/y

Anyakora, Engel and Lees give for control valves 0.6 failures/y (Lees, p.343) and found that of 188 cases where failure mode was recorded 54 (29%) (Lees, p.346) were leakages, though this could have included a few internal leakages. The overall failure rates are probably on the high side.

There is a distinction to be made between control valves and manual valves. UKAEA data (Lees, p.342) give failure rates of 0.25 failures/y for control valves and 0.13 failures/y for hand valves. The

Rijnmond Report (p.375) give 0.3 and 0.1 failures/y for control and hand valves, respectively.

Aupied et al. (1983) obtained following data on valve failures:

No. of valves = 1022

No. of failures = 314

Observation period = 9315 h = 1.063 y

Hence

Valve failure frequency = 315/(1022 × 1.063) = 0.29 failures/y

They also found for proportion of failures which involved external leakage

Proportion of failures involving leaks = 0.3

Hence

Valve leak frequency = 0.3 × 0.29 = 0.09 leaks/y

The valves studied by Aupied et al. appear to be a mixture of control and hand valves. Valves include gate as well as globe valves and drain as well as control valves. Assuming their data apply both to control and to hand valves, the latter would almost certainly be more numerous so that the failure rate would be biased towards the value for hand valves.

The Rasmussen Report gives for valve rupture frequency

Valve rupture frequency = 1×10^{-8}/h = 0.9×10^{-4}/y

It is not clear precisely what constitutes rupture.

On the basis of the foregoing, the valve leak frequencies adopted are as follows:

Valve leak frequency: rupture = 10^{-5} leaks/y

Valve leak frequency: major leak = 10^{-4} leaks/y

Valve leak frequency: minor leak = 10^{-3} leaks/y

The area A of a rupture leak is taken to be the cross-sectional area of the pipe and the areas for major and minor leaks are taken to be 0.1A and 0.01A. No distinction is made between hand and control valves.

A8.4   PUMP FAILURE AND LEAK FREQUENCY

Information on pump failure is given by the following sources:

Nixon (1971)
Nature of pump leaks, p.153

Rasmussen Report (1974)
Pump failure frequency – see Lees (1980), Table A9.3

Kletz (1977)
Pump leak frequencies, p.52

First Canvey Report (1978)
Pump leak frequencies – see Lees (1980), p.1021 and Table A10.3

Wallace (1979)
Pump seal leak frequencies, Table VIII

Welker and Schorr (1979)
LNG pump failure frequencies

Sherwin and Lees (1980)
Pump failure frequencies and modes

Rijnmond Report (1982)
Pump leak frequencies, Table IX.1, p.374

Summers-Smith (1982)
Pump seal leak rates, p.7/3

Aupied, Le Coguiec and Procaccia (1983)
Pump failure frequencies, Table 7
Proportion of packing failures, p.144

Sherwin (1983)
Pump failure frequencies, p.207 and Table 5
Pump seal failures and leaks as proportion of pump failures, Table 1

Anon (1985)
Pump failure frequency

     The data given in the above sources include the following. Wallace gives
Pump seal leak frequency $= 0.12$ leaks/y
     Aupied *et al.* obtained the following data on pump failures:
Pump failure frequency $= 5.6 \times 10^{-4}$ failures/h $= 4.9$ failures/y
Proportion of packing failures $= 0.3$
They studied only 20 boiler feedwater pumps.
     Sherwin studied 85 ethylene plant pumps and obtained the following data:
Observation period $= 19$ months $= 1.58$ y
No. of pumps $= 85$

No. of failures = 243

Pump failure frequency = $243/(85 \times 1.58)$ = 1.8 failures/y

Proportion of gland/seal failures = $119/243$ = 0.49

Proportion of leaks = 0.30

Pump leak frequency = 0.30×1.8 = 0.54 leaks/y

Welker and Schorr give for LNG pumps

MTBF = 3500 h

Failure frequency = 2.5 failures/y

Anon. (1985) states that ANSI standard pump failure rate has

MTBF = 6 months

The Rasmussen Report gives for pumps (failure to run)

Failure frequency = 3×10^{-5}/h = 0.26 failures/y

This is a rather low value compared with the others.

The Rijnmond Report gives for failure frequency of pumps the same value as the Rasmussen Report.

The Rijnmond Report gives for pump catastrophic failure

Pump catastrophic failure frequency = 1×10^{-4} failures/y

On the basis of the foregoing, the pump leak frequencies are estimated as follows:

Frequency of pump failure = 2 failures/y

Proportion of leaks = 0.30

Pump leak frequency = 0.30×2 = 0.6 leaks/y

This is for all leaks. For rupture leaks

Frequency of pump rupture = 1×10^{-4} failures/y

From these estimates the following pump leak frequencies have been adopted:

Frequency of pump leak: rupture leak = 3×10^{-5} leaks/y

Frequency of pump leak: major leak = 3×10^{-4} leaks/y

Frequency of pump leak: minor leak = 3×10^{-3} leaks/y

The area A of a rupture leak is taken to be the cross-sectional area of the pipe and the areas for major and minor leaks are taken to be 0.1A and 0.01A.

Information on pump fires and vapour cloud explosions is also available:

Kletz (1971)

Pump fire frequencies, Table IV

Kletz (1977)

Vapour cloud explosion frequencies, p.52

For pump fires Kletz (1971) gives the estimates shown in Table A8.5.

Table A8.5   Estimated frequencies of pump fires (Kletz, 1977)

Material in pump	Estimated frequency of fires (fires/pump y)
LPG below ambient temperature	$>10^{-3}$
Material above autoignition temperature	$>10^{-3}$
Flammable material well above ambient temperature (eg petrol at 70°C, gas oil at 150°C)	$<10^{-3}$ $>10^{-4}$
LPG at ambient temperature	$<0.33 \times 10^{-3}$ $>0.33 \times 10^{-4}$
Petrol or similar materials at or near ambient temperature	$<10^{-4}$ $>10^{-5}$
Gas oil or similar materials below their flash points	$<10^{-5}$

For vapour cloud explosions on a petrochemicals plant Kletz (1977) gives the following estimates:

VCE frequency for normal pumps $= 10^{-2}$ VCEs/plant y

VCE frequency for severe duty pumps $= 10^{-1}$ VCEs/plant y

A8.5   COMPRESSOR FAILURE AND LEAK FREQUENCY

Information on reciprocating compressor leak frequency is given by:

Kletz (1977)

Compressor leak frequencies, p.52

For vapour cloud explosions on a petrochemicals plant from leaks on reciprocating compressors Kletz (1977) gives the following estimates:

VCE frequency for reciprocating compressors $= 0.1$ VCEs/plant y

A8.6   PRESSURE RELIEF VALVE FAILURE AND LEAK FREQUENCY

Information on failure and leak frequency of pressure relief valves is given by:

Lawley and Kletz (1975)

PRV lift light frequencies – see Lees (1980), Table on p.1009

Wallace (1979)
Safety relief valve leak frequencies, Table VIII

Aird (1982)
PRV failure frequencies

Rijnmond (1982)
PRV lift light frequencies, Table IX.1, p.378

The data given in the above sources include the following. Wallace gives

Safety relief valve leak frequency = 0.07 leaks/y

A8.7 SMALL BORE CONNECTION FAILURE AND LEAK FREQUENCY

There is little information on failure of small bore connections as such, but an estimate may be made from data given for pneumatic connections, mainly on instruments. Lawley and Kletz (1975) give for fracture of air supply lines:

Frequency of air line fracture = 0.01 failures/y

They give the same frequency for blockage. For failure of pneumatic connectors Smith (1988) gives:

Frequency of pneumatic connector failure = 0.013 failures/y

This presumably includes both fracture and blockage. Allowing for a proportion of connections which are somewhat more robust, the following small bore leak frequencies have been adopted:

Frequency of small bore connection leak: rupture leak = 0.0005 failures/y
Frequency of small bore connection leak: major leak = 0.005 failures/y

A8.8 LEAK FREQUENCY FOR FUGITIVE EMISSIONS

Although the same components such as valves and pumps are involved here, the leak flow rates tend to be much lower. For this reason they are treated separately.

Bochinski and Schultz (1979)
List of leak sources on chemical plants, Tables 1 and 2

Freeberg and Aarni (1982)
Contribution of different leak sources to total refinery emission, with total leak rates from each source, Table 2
Proportion of flanges, valves leaking, Table 3

Hughes et cl. (1979)
Leak rates for components, Tables 2 and 4

Jones (1984)
Contribution of different leak sources to total emission, with total leak rates from each source (EPA estimates), Table 1
Leak rate distribution for valves, Table 2
Leak rates for components (EPA estimates), Table 3

Kunstman (1980)
Leak rate categories, 2nd table
Leak rates for components before maintenance, 3rd table

Morgester et al. (1979)
Proportion of flanges and valves leaking, Table 2
Valve leak frequency and rates, Table 3

Rosebrook (1977)
Leak rates for components, Table 1

Wallace (1979)
Leak rates for components (EPA standards), Table 4
Leak rates for components (comparative review), Table V
Leak frequency for components, Table VIII

Wetherold et al. (1981)
Leak rates for components and 95% confidence limits on leak rates, Table 1

Wetherold et al. (1983)
Proportion of valves leaking and leak rates, Table 1

APPENDIX 9—MODELS FOR EMISSION AND DISPERSION

The set of models used for emission and dispersion is given without comment in Section 12. In this appendix comments are given and in some cases alternative models are discussed.

The notation used is given at the end of the appendix.

A9.1 REVIEW OF MODELS SELECTED

EMISSION

Gas/vapour
Equations (12.1) – (12.4) are those in general use and require no comment.

Liquid
Equation (12.5) also is that in general use.

Coefficient of discharge
Again equations (12.6a) – (12.6c) are those in general use.

Two-phase flashing flow
Two-phase flow is extremely complex and there are a large number of treatments available.

Two-phase flow through an orifice is metastable and the liquid flow equation may be used. As the length of the outlet is increased, flashing starts to occur so that for such flow a two-phase flow correlation is required.

A treatment by Fauske (1965) has been widely used in industry, for hazard assessement and other applications (eg Simpson, 1971). The basic equation is

$$G = C_D\sqrt{(2\rho_L(P_1 - P_c))} \qquad \text{(A9.1)}$$

where P_c is the critical exit pressure (N/m^2). Fauske presents a graph in which P_c is correlated against the outlet length/diameter ratio l/d.

There are also refinements of Fauske's equation (eg Jones and Underwood, 1983).

Subsequent work has shown that the correlator for flow should be the length rather than the length/diameter ratio. This is demonstrated by the work of Fletcher (1984) and of other workers and is accepted by Fauske (1985).

Equation (A9.1) has been discussed by Fletcher and Johnson (1984). They draw attention to the difficulty of determining P_c. They are able in fact to correlate their own results and those of other workers by plotting

$$\frac{G^2}{C_D^2 P_1} \text{ versus } l$$

Their plot for saturated water is shown in Figure A9.1. The intercept on the vertical axis corresponds to zero pipe length, or an orifice, and is about 2000, which would correspond to $2\rho_L$.

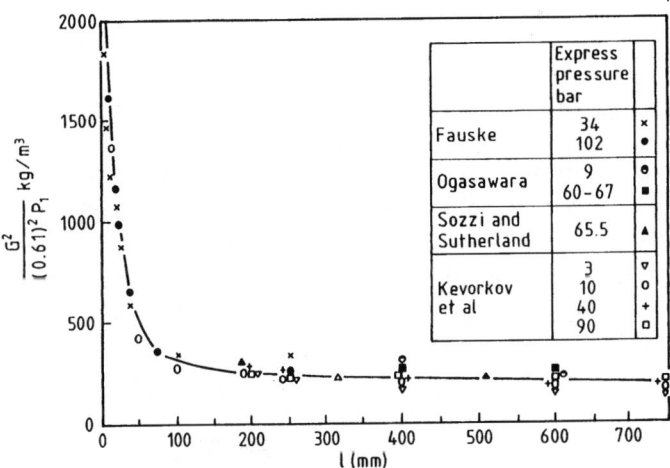

Figure A9.1 Correlation for two-phase flow in a short pipe section (after Fletcher and Johnson, 1984)

This approach has been adopted in the present work resulting in equation (12.8). The function $f(l)$ given in Figure 12.1 has been derived from Figure A9.1.

147

Some deviation from the function given may be expected for fluids with a density different from that water, but the data given by Fletcher and Johnson for Refrigerant 11 suggest that this is much less important than the effect of pipe length and the correlation of equation (12.8) is regarded as sufficiently accurate for present purposes.

FLASHING LIQUIDS

Flash fraction
Equation (12.9) is that in general use.

Spray fraction
Equations (12.10) and (12.11) are simply an encoding of the rule-of-thumb that the amount of spray formed is approximately equal to the amount of vapour and that if half the liquid flashes off as vapour virtually all the liquid is then either spray of vapour.

Since the formulation of this rule-of-thumb more has been learned about drop formation and rainout of drops, but no correlation for spray formation has been found.

VAPORISATION OF LIQUIDS
The equations given for pool area and for vaporisation are for continuous leaks.

An equation for pool area is given by Mecklenburgh (1985). The same equation is given for both water and land, but with a correction factor δ, which is unity for land. This equation appears to be derived from the equation obtained by Shaw and Briscoe (1978) for pool size on water and to have been adapted for use on land. Equation (12.12a) is based on the equation derived by Shaw and Briscoe for land.

Equation (12.12b) is that given by Mecklenburgh and the values of pool depth h_m are those given by him.

Equations (12.13) and (12.14) and the associated constants are those given by Mecklenburgh. They are evidently derived from the equation for evaporation from a pool given by Sutton (1934) as adapted by Pasquill (1943).

These equations apply to a pool at constant temperature. In due course the pool will chill due to the vaporisation. A treatment for this case has been given by Mackay and Matsugu (1973).

Equation (12.13) is therefore applicable only to the initial spill before chilling, but this is the worst case, in that chilling will reduce the rate of vaporisation.

Vaporization of a cryogenic liquid

Equation (12.15) is that in general use. It is readily derived theoretically omitting the surface correction factor.

The equation is now frequently quoted with the surface correction factor (eg Shaw and Briscoe, 1978; Mecklenburgh, 1985). The values of the surface correction factor given are those of Mecklenburgh.

DISPERSION

Gas/vapour jet

Most of the equations given in the literature are for subsonic rather than sonic, or underexpanded, jets.

Equation (12.17) is that applicable to a subsonic circular jet. It derives from the work of Ricou and Spalding (1961) and of Long (1963).

Other configurations for which relations are available include the plane jet and the radial jet. A plane jet is a jet issuing from a slot and a radial jet one issuing from a circumferential slit opening in a cylinder. The latter is exemplified by a jet from a gasket burst on a pipe flange. Equations for plane and radial jets are given by Rajaratnam (1976) and Chen and Rodi (1980).

The equations for circular, plane and radial jets are given at the end of this subsection.

In general, relations tend to be more readily available for the axial velocity ratio than for the axial concentration ratio. Further, the axial concentration ratio tends to be somewhat less than the axial velocity ratio. For a circular jet the ratio of these two ratios is $5/6.2 = 0.8$.

Separate treatment is required for the sonic, or underexpanded, jet. Equation (12.17) and the suppporting equations for a circular jet are based on the work of Ewan and Moodie (1986).

Liquid jet

Equation (12.24) is the standard relation for a projectile in the absence of air drag and disintegration derived from ballistic theory.

A relation may also be derived theoretically for the case where the air resistance is proportional to the velocity (eg Brown, 1969).

Disintegration of jets has also been studied and expressions are available for the estimation of the distance along the jet at which breakup starts to occur (eg Engh and Larsen, 1979).

This does not itself, however, allow the jet travel to be estimated and no expression has been found for this.

Equation (12.25) encodes the rule-for-thumb quoted by O'Shea (1982) that in practice a liquid jet travels about half the distance predicted theoretically.

Gas dispersion

Gas dispersion also is extremely complex and again there are a large number of treatments available. In particular, there are separate treatments for heavy gas dispersion.

Most treatments of gas dispersion, including those for heavy gases, apply to a source with low momentum. While models are available for high velocity gas jets, there is little on dispersion of flashing vapour from high velocity liquid jets.

Equation (12.24) is the equation for a continuous point source in the Pasquill-Gifford model for neutral density gas, which is widely used. Another treatment of dispersion of neutral density gas over short ranges is that of Katan (1951).

Peak/mean concentration ratios

An account of peak/mean concentration ratios for jets and plumes is given by Long (1963). For jets he gives equations for both time mean and instantaneous axial concentrations. The ratio of the latter to the former is 1.5 ($= 9/6$).

Long states that for a passive plume the ratio is unlikely to exceed 2. It seems probable that the ratio for a plume will be somewhat greater than for a jet, which has a relatively well defined shape. The value of 2, which has been widely used, is adopted here also.

Circular, planar and radial gas jets

Equations for the velocity on the centreline of circular, planar and radial jets are given by Rajaratnam (1976) and for both the centreline velocity and concentration of circular and planar jets by Chen and Rodi (1980).

An equation for the centreline concentration of a radial jet may be may be inferred from these relations.

Circular jet

For the concentration on the centreline of the jet

$$\frac{c_{cl}}{c_o} = 5\left(\frac{\rho_\alpha}{\rho_o}\right)^{0.5}\frac{d_o}{x}$$

(A9.2)

and for the gas velocity on the centreline of the jet

$$\frac{u_{cl}}{u_o} = 6.2\left(\frac{\rho_\alpha}{\rho_o}\right)^{-0.5}\frac{d_o}{x}$$

(A9.3)

Planar jet

For the concentration on the centreline of the jet

$$\frac{c_{cl}}{c_o} = 2\left(\frac{\rho_\alpha}{\rho_o}\right)^{0.5}\left(\frac{2b_o}{x}\right)^{0.5}$$

(A9.4)

and for the gas velocity on the centreline of the jet

$$\frac{u_{cl}}{u_o} = 2.4\left(\frac{\rho_\alpha}{\rho_o}\right)^{-0.5}\left(\frac{2b_o}{x}\right)^{0.5}$$

(A9.5)

Radial jet

For the concentration on the centreline of the jet

$$\frac{c_{cl}}{c_o} = 2.8\left(\frac{\rho_\alpha}{\rho_o}\right)^{0.5}\frac{r_o}{r}$$

(A9.6)

and for the gas velocity on the centreline of the jet

$$\frac{u_{cl}}{u_o} = 3.5\left(\frac{\rho_\alpha}{\rho_o}\right)^{-0.5}\frac{r_o}{r}$$

(A9.7)

A9.2 APPLICABILITY OF DISPERSION MODELS OVER SHORT RANGES

The ranges involved in hazardous area work are relatively short. It is necessary, therefore, to use models which are applicable over short ranges. This applies particularly to models for gas dispersion, where the principal models are typically used over ranges of hundreds or thousands of metres.

The classical neutral density gas dispersion models are the Sutton and Pasquill-Gifford models, the latter being derived from the former. As it happens, the question of the range of his model was specifically considered by Sutton (1950). He describes experiments on the dispersion of a vertical jet of hot gas for which he obtained a value of the diffusion coefficient C. This value is similar to that obtained in the dispersion of smoke from a smoke generator over a distance of 100 m, provided that for comparability use is made in the latter case of the instantaneous rather than the time width of the plume. Sutton states

"This leads to the striking conclusion that the same coefficient of diffusion is valid for the spread of smoke over hundreds of meters as for the mixing of a tiny column of hot air with the cold air of a laboratory."

The use of the Sutton model for estimation of dispersion in relation to hazard areas around vents has been treated by Long (1963), who devotes an appendix to the applicability of the model over short ranges. He discusses both the basic assumptions and available experimental results. He reviews the assumptions systematically and finds no fundamental inapplicability for short ranges. In particular, the semi-angle of divergence of the roughly conical plume predicted using the values given by Sutton for the diffusion coefficient is about 10° and 16° in the vertical and horizontal directions, respectively, which compares well with semi-angles in the range 9–16° found for jets and plumes. The only experimental work found by Long was that of Katan (1951). Using values of the Sutton coefficient based on a three-minute sampling time, Long found that the Sutton equations tend to overestimate the concentration somewhat.

It is concluded from the above that the conventional neutral density gas dispersion models are sufficiently accurate over short ranges for present purposes.

A9.3 APPLICABILITY OF DISPERSION MODELS IN LOW WIND CONDITIONS

It is also necessary to consider the problem of dispersion in near calm, or low wind, conditions. Outdoors these occur typically on clear nights with frost or heavy dew.

Conditions which are usually classified a Pasquill category F apply in the UK for some 20% of the time. Included in these are period of calm. Beattie (1961) found that the proportion of time for which such conditions apply is about 5 to 8%. Pasquill actually excluded such conditions from his original categories, because dispersion then tends to be irregular. Beattie has classified such conditions as category G, but he does not give the associated dispersion coefficients.

It is natural to assume that dispersion is less in G than in F conditions. This assumption is made, for example, in NRC Regulatory Guide 1.21.

However, there is evidence to show that under G conditions dispersion is ill-defined. Work on such conditions by Sagendorf (1975) has shown that a plume is subject to considerable meander, while work by van der Hoven (1976) has indicated values of dispersion coefficients which correspond to those found across the whole range from A to F.

It would appear, therefore, that the use of Pasquill F values for the dispersion coefficients would be a not unreasonable approach and would probably tend to be a conservative one, but dispersion beyond the envelope so determined should be expected in a proportion of cases.

NOTATION

b_o halfwidth (plane aperture, radial aperture) (m)

c concentration (kg/m^3)

C_D coefficient of discharge

d_o diameter of pipe (or circular aperture) (m)

G mass flow per unit area (kg/m^2 s)

k expansion index

l length of pipe (m)

P pressure (N/m^2)

r radial distance (m)

r_o radius of radial aperture (m)

u velocity (m/s)

x distance from exit (m)

ρ density (kg/m^3)

Subscripts

a air

c critical

cl centreline

L liquid

o exit conditions

1   upstream

2   downstream

APPENDIX 10—PHYSICAL PROPERTIES OF SELECTED SUBSTANCES

This appendix contains the equations used for the correlation of physical properties and the physical property data for the six representative fluids.

Some basic physical properties of these fluids are given in Table A10.1.

Table A10.1 Physical properties of representative fluids

	Hydrogen	Methane	Ethylene	Propane	Acetone	Methyl ethyl ketone
Molecular weight	2	16	28.05	44.08	58.08	72.11
Normal boiling point (K)	20.28	112	169.3	231.1	329.2	352.6
Liquid density (at BPt) (kg/m^3)	70	424	569.6	582	789.9[a]	805.4[a]
Latent heat of vaporisation (kJ/kg)	451.9	577.4	515.2	458	550.5	473.5
Critical pressure (bar)	12.9	46.3	50.5	42.5	47	42.1
Critical temperature (K)	33.25	190.65	282.9	369.6	508.5	537
Critical volume (cm^3/mol)	64.3	99.2	130.4	203	209	267
Acentric factor	−0.216	0.011	0.089	0.153	0.304	0.320
Dipole moment (debye)	0	0	0	0	2.9	3.3
Polar correction factor κ	n/a	n/a	0	n/a	n/a	n/a
Lower flammability limit (% v/v)	4.0	4.9	2.7	2.0	2.15	1.8
Upper flammability limit (% v/v)	75.6	15.9	34	9.5	13	11.5
C_{pL} (at 298 K) (kJ/kg K)					1.652	2.176	2.211

Notes:

(a) At 293 K.

The notation used is given at the end of the appendix.

Two types of equation are given: equations for prediction where there are no data available and equations for correlation of available data points.

VAPOUR PRESSURE

The equation used is that of Wagner (1973):

$$\ln P_{vr} = \frac{At + Bt^{1.5} + Ct^3 + Dt^6}{T_r} \qquad (A10.1)$$

with

$$P_{vr} = \frac{P_r}{P_c} \qquad (A10.2)$$

$$t = (1 - T_r) \qquad (A10.3)$$

$$T_r = \frac{T}{T_c} \qquad (A10.4)$$

The constants A–D are given by Reid, Prausnitz and Poling (1987) and are listed in Table A10.2.

Table A10.2   Physical properties of representative fluids: coefficients for vapour pressure equation

	Hydrogen	Methane	Ethylene	Propane	Acetone	Methyl ethyl ketone
A	−5.57929	−6.00435	−6.32055	−6.72219	−7.45514	−7.71476
B	2.60012	1.18850	1.16819	1.33236	1.20200	1.71061
C	−0.85506	−0.83408	−1.55935	−2.13868	−2.43926	−3.68770
D	1.70503	−1.22833	−1.83552	−1.38551	−3.35590	−0.75169
Range(K)[a]	14	91	105	145	259	255

Notes:

(a) Range is from temperature given to critical temperature.

LIQUID DENSITY

The equations used are those of Rackett (1970). For prediction

$$V_s = \left(\frac{R'T_c}{P_c}\right) Z_{RA}^{1 + (1 - T_r)^{2/7}} \tag{A10.5}$$

with

$$Z_{RA} = 0.29056 - 0.08775\omega \tag{A10.6}$$

For correlation

$$V_s = V_s^R (Z_{RA})^J \tag{A10.7}$$

with

$$J = (1 - T_r)^{2/7} - (1 - T_r^R)^{2/7} \tag{A10.8}$$

LIQUID SPECIFIC HEAT

The equation used for correlation is that of Bondi and Rowlinson (1969):

$$\frac{(C_{pL} - C_p^o)}{R} = 1.45 + 0.45(1 - T_r)^{-1}$$

$$+ 0.25\omega[17.11 + 25.2(1 - T_r)^{1/3} T_r^{-1} + 1.742(1 - T_r)^{-1}] \tag{A10.9}$$

with

$$\frac{(C_{pL} - C_{dL})}{R} = \exp(20.1T_r - 17.9) \tag{A10.10}$$

$$\frac{(C_{dL} - C_{sat.L})}{R} = \exp(8.655T_r - 8.385) \tag{A10.11}$$

LIQUID VISCOSITY

The equation used is that given by Reid, Prausnitz and Poling (1987):

$$\ln\eta = A + (B/T) + CT + DT^2 \tag{A10.12}$$

The constants A–D are given by these authors and are listed in Table A10.3.

Table A10.3   Physical properties of representative fluids: coefficients for liquid viscosity equation

	Hydrogen	Methane	Ethylene	Propane	Acetone	Methyl ethyl ketone
A		-26.87	-17.74	-7.764	-4.0033	-4.213
B		1150	1078	721.9	845.6	975.9
C		0.1871	0.08577	2.381×10^{-2}		
D		-5.211×10^{-4}	-1.758×10^{-4}	-4.665×10^{-5}		
Reference value (cP)		0.14	0.031	0.091	0.32	0.42
Reference temperature (K)		103	273	298	298	294
Range (lower)(K)		93	104	86	193	273
(upper)(K)		357	282	369	333	353

GAS DENSITY

The ideal gas law is assumed.

GAS SPECIFIC HEAT

The equation used for correlation is that given by Reid, Prausnitz and Poling:

$$C_p = A + BT + CT^2 + DT^3 \tag{A10.13}$$

The constants A–D are given by these authors and are listed in Table A10.4.

Also for the ratio of specific heats γ:

$$\gamma = \frac{C_p}{C_p - R} \tag{A10.14}$$

Table A10.4 Physical properties of representative fluids: coefficients for gas specific heat equation

	Hydrogen	Methane	Ethylene	Propane	Acetone	Methyl ethyl ketone
A	27.14	19.25	3.806	−4.224	6.301	10.94
B	9.274 $\times 10^{-3}$	5.213 $\times 10^{-2}$	0.1566	0.3063	0.2606	0.3559
C	−1.381 $\times 10^{-5}$	1.197 $\times 10^{-5}$	−8.348 $\times 10^{-5}$	−1.586 $\times 10^{-4}$	−1.253 $\times 10^{-4}$	−1.9 $\times 10^{-4}$
D	7.645 $\times 10^{-9}$	-1.132 $\times 10^{-8}$	1.755 $\times 10^{-8}$	3.215 $\times 10^{-8}$	2.303 $\times 10^{-8}$	3.920 $\times 10^{-8}$

GAS VISCOSITY

The equation used for prediction is that of Chung, Lee and Starling (1984):

$$\eta = 40.785 \left[\frac{F_c(MT)^{1/2}}{V_c^{2/3}\Omega_v} \right] \tag{A10.15}$$

with

$$\Omega_v = [A(T^*)^{-B}] + C[\exp(-DT^*)] + E[\exp(-FT^*)] \tag{A10.16}$$

$$F_c = 1 - 0.275\omega + 0.059035\mu_r^4 + \kappa \tag{A10.17}$$

$$\mu_r = 131.3 \left(\frac{\varphi}{(V_c T_c)^{1/2}} \right) \tag{A10.18}$$

$$T^* = 1.2593T_r \tag{A10.19}$$

The values of the constants A–F are

A	1.16145	D	0.77320
B	0.14874	E	2.16178
C	0.52487	F	2.43787

NOTATION

A–F constants (specific to the particular property)

C_{dl} change of enthalpy of saturated liquid with temperature (kJ/kmol K)

C_p specific heat of gas (kJ/kmol K)

C_p^o specific heat of ideal gas at T_r (kJ/kmol K)

C_{pl} specific heat of liquid at constant pressure (kJ/kmol K)

$C_{sat,1}$ energy required to effect a temperature change whilst keeping the liquid saturated (kJ/kmol K)

F_c shape and polarity factor

M molecular weight

P_c critical pressure (bar)

P_v vapour pressure (bar)

P_{vr} reduced vapour pressure

R universal gas constant ($= 83.14$ bar cm^3/(mol K) in equation (A10.7); 8.314 kJ/kmol K in equations (A10.9)–(A10.11), (A10.14))

t temperature function defined by equation (A10.3)

T absolute temperature (K)

T_c critical temperature (K)

T_r reduced temperature

T^* function defined by equation (A10.19)

V_c critical volume (cm^3/*mol*)

V_s saturated liquid volume (cm^3/*mol*)

Z_{RA} function defined by equation (A10.6)

Greek letters

γ ratio of specific heats of gas

η liquid viscosity (cP) (equation (A10.12))

η gas viscosity (micropoise) (equation (A10.15))

κ polar correction factor for highly polar molecules

μ dipole moment (debye)

μ_r dimensionless dipole moment

ω acentric factor

Ω_v viscosity collision integral

Subscript

R reference conditions

APPENDIX 11—NUMERICAL INVESTIGATION OF HAZARD RANGES

A numerical investigation has been carried out to determine the hazard ranges for a number of leak scenarios. These scenarios and the corresponding hazard ranges obtained are shown in Tables A11.1 – A11.6.

The mass flows have been calculated using the appropriate gas or liquid flow equations, the flashing liquid flows of propane in Tables A11.3, A11.5 and A11.6 being determined from the two-phase flow correlation. For the dispersion of gas in Table A11.1 and of total flash in Table A11.3 the circular, underexpanded jet model has been used. For dispersion of non-flashing liquids in Table A11.2 two cases are shown. In Section B the dispersion has been calculated assuming that the liquid forms a fine spray and vaporises completely and then using the Gaussian plume dispersion model and in Section C assuming that the liquid jet is maintained and thus using the liquid jet model. For dispersion of flashing liquids in Table A11.3 it has been assumed that all the liquid flashes off or forms spray and that the leak momentum is high so that the circular, underexpanded jet model is applicable. For the dispersion of non-flashing liquids in Table A11.5 it has been assumed that the leak momentum is low, the liquid is held within a bund of area 1 m^2 and the liquid vaporisation model has been used, followed by the Gaussian plume model, while for the flashing liquid in this table it has been assumed that again that the leak momentum is low, that all the liquid flashes or forms spray and the Gaussian plume model has been used. For the dispersion of flashing liquid in Table A11.6 it has been assumed that the momentum is low, the total flash, vapour and spray, has been estimated and the Gaussian plume model used.

This investigation is considered in Section 14.

Table A11.1 Flows and hazard ranges for flange leaks: gas

A Flow (kg/s)

Pressure (*bar*)	10			20		
Hole size (*mm*^{2})	0.25	2.5	25	0.25	2.5	25
Fluid						
Hydrogen	9.2×10^{-5}	9.2×10^{-4}	0.0092	1.8×10^{-5}	1.8×10^{-4}	0.0018
Methane	2.6×10^{-4}	2.6×10^{-3}	0.026	5.2×10^{-4}	5.2×10^{-3}	0.052
Ethylene	3.4×10^{-4}	3.4×10^{-3}	0.034	7.0×10^{-4}	7.0×10^{-3}	0.070

B Distance to LFL (m)

Pressure (*bar*)	10			20		
Hole size (*mm*^{2})	0.25	2.5	25	0.25	2.5	25
Fluid						
Hydrogen	0.62	2.0	6.2	0.88	2.8	8.8
Methane	0.19	0.60	1.9	0.26	0.84	2.7
Ethylene	0.26	0.84	2.7	0.39	1.2	3.9

C Distance to 0.67 LFL (m)

Pressure (*bar*)	10			20		
Hole size (*mm*^{2})	0.25	2.5	25	0.25	2.5	25
Fluid						
Hydrogen	0.94	3.0	9.4	1.3	4.3	13
Methane	0.28	0.89	2.8	0.40	1.3	4.0
Ethylene	0.30	1.3	3.9	0.58	1.9	5.8

Notes:

(a) It is assumed that gas disperses as a circular, underexpanded jet.

Table A11.2 Flows and hazard ranges for flange leaks: liquid

A Flow (kg/s)

Pressure (bar)	10			20		
Hole size (mm^2)	0.25	2.5	25	0.25	2.5	25
Fluid						
Acetone	6.0×10^{-3}	0.060	0.60	8.4×10^{-3}	0.084	0.84
Methyl ethyl ketone	6.0×10^{-3}	0.060	0.60	8.5×10^{-3}	0.085	0.85

B Distance from 25 mm^2 hole to LFL, assuming liquid forms fine spray (m)[a]

Pressure (bar)	10	20
Fluid		
Acetone	25	
Methyl ethyl ketone	25	

C Distance to point of impact, assuming liquid remains as jet (m)

Pressure (bar)	10	20
Fluid		
Acetone	47	94
Methyl ethyl ketone	46	92

Notes:

(a) A check on the drop size of the liquid jet may be made using the method given by Wheatley (1987). For both these liquids the drop size is less then 0.05 mm. Given this small drop size, it is assumed that the drops also vaporise and that the resultant vapour disperses as a Gaussian plume.

Table A11.3 Flows and hazard ranges for flange leaks: flashing liquid

A Flow (kg/s)

Pressure (bar)	10			20		
Hole size (mm^2)	0.25	2.5	25	0.25	2.5	25
Fluid						
Propane	4.7×10^{-3}	0.047	0.47	6.6×10^{-3}	0.066	0.66

B Distance to LFL (m)

Pressure (bar)	10			20		
Hole size (mm^2)	0.25	2.5	25	0.25	2.5	25
Fluid						
Propane	0.30	0.96	3.0	0.44	1.4	4.4

C Distance to 0.67 LFL (m)

Pressure (bar)	10			20		
Hole size (mm^2)	0.25	2.5	25	0.25	2.5	25
Fluid						
Propane	0.45	1.5	4.5	0.67	2.3	6.7

Notes:

(a) It is assumed that the total flash (vapour and spray) is unity and that the resultant vapour disperses as a circular, underexpanded jet.

Table A11.4   Flows and hazard ranges for compressor leaks

A Flow (kg/s)

Pressure (bar)	10		20	
Hole size (mm²)	25	250	25	250
Fluid				
Hydrogen	0.0092	0.092	0.0184	0.184
Methane	0.026	0.26	0.052	0.52
Ethylene	0.034	0.34	0.070	0.70

B Distance to LFL (m)

Pressure (bar)	10		20	
Hole size (mm²)	25	250	25	250
Fluid				
Hydrogen	8.9	28	13	40
Methane	4.8	15	6.7	21
Ethylene	5.6	18	8.1	25

C Distance to 0.5 LFL (m)

Pressure (bar)	10		20	
Hole size (mm²)	25	250	25	250
Fluid				
Hydrogen	13	40	18	57
Methane	6.8	22	9.7	31
Ethylene	8.0	25	12	36

Notes:

(a) It is assumed that the momentum of the emission is low and that the gas disperses as a Gaussian plume.

Table A11.5 Flows and hazard ranges for pump leaks

A Flow (kg/s)

Pressure (bar)	10			20		
Hole size (mm^2)	0.25	2.5	25	0.25	2.5	25
Fluid						
Acetone	6.0×10^{-3}	0.060	0.60	8.4×10^{-3}	0.084	0.84
Methyl ethyl ketone	6.0×10^{-3}	0.060	0.60	8.5×10^{-3}	0.085	0.85

B Distance to LFL (m)

Pressure (bar)	10			20		
Hole size (mm^2)	0.25	2.5	25	0.25	2.5	25
Fluid						
Acetone	2.1 all cases					
MEK	1.3 all cases					
Propane	1.9	6.1	19	2.3	7.3	22

C Distance to 0.5 LFL (m)

Pressure (bar)	10			20		
Hole size (mm^2)	0.25	2.5	25	0.25	2.5	25
Fluid						
Acetone	2.9 all cases					
MEK	1.8 all cases					
Propane	2.8	8.7	27	3.2	10	33

Notes:

(a) It is assumed that the momentum of the emission is low, that the liquid is caught in a bund of area 1 m^2 and that the vapour from the pool disperses as a Gaussian plume.

(b) It is assumed that the momentum of the emission is low, that the total flash (vapour and spray) is unity and that the resultant vapour disperses as a Gaussian plume.

Table A11.6 Flows and hazard ranges for drain/sample point leaks

Hole size: 15 mm diameter drain/sample pipe 500 mm long

A Flow (kg/s)		
Pressure (bar)	*10*	*20*
Propane	1.5	2.1
B Distance to LFL (m)		
Pressure (bar)	*10*	*20*
Fluid		
Propane	22	26
C Distance to 0.5 LFL (m)		
Pressure (bar)	*10*	*20*
Fluid		
Propane	31	37

Notes:

(a) It is assumed that emission is two-phase, that the total flash fraction (vapour and spray) is 0.81 and that the resultant vapour disperses as a Gaussian plume.

APPENDIX 12—LEAKS FROM A PIPE FLANGE

The emission and dispersion of material from leaks from a pipe gasket failure has been investigated as an illustration of the ranges which may be expected from a particular equipment leak.

The following scenarios have been considered:

Leak hole size

a) Planar hole, 25 mm long × 1 mm wide

b) Circular leak, same equivalent diameter as (a)

The process fluids and operating conditions conditions considered are shown in Table A12.1.

The emissions have been calculated using the standard equations for gas flow, subsonic or sonic, and for liquid flow. The liquid flow equations have been used for flashing fluids. A coefficient of discharge of 0.6 has been used in each case.

For dispersion of gas jets the equations used are those given in Section 12 and Appendix 9.

For a non-flashing liquid jet two estimates are made. For the first air resistance is neglected. For this case

$$x = \frac{u^2}{g}\sin2\alpha \qquad (A12.1)$$

For the second case where air resistance is taken into account

$$x = \frac{u\cos\alpha}{k}[1 - \exp(-kt)] \qquad (A12.2)$$

Table A12.1 Dispersion of fluids from a pipe gasket leak

Case	Fluid	Pressure (bar a)	Temperature (°C)	Distance to LFL (m)
1 Subsonic jet, circular	Methane, gas	2	20	0.78
2 Subsonic jet, plane	Methane, gas	2	20	3.0
3 Sonic jet, circular	Methane, gas	5	20	1.3
4 Sonic jet, circular	Methane, gas	10	20	1.9
5 Sonic jet, circular	Propane, gas	5	20	2.0
6 Sonic jet, circular	Propane, gas	10	20	3.0
7 Liquid jet, no resistance[a][b]	Acetone, liquid	5	20	55
8 Liquid jet, no resistance[c]	Acetone, liquid	10	10	107
9 Liquid jet, resistance[d]	Acetone, liquid	5	20	37
10 Liquid jet, resistance[e]	Acetone, liquid	10	20	60
11 Liquid jet[f]	Propane, liquid	10	20	26

Notes:

(a) For cases 7–10 it is assumed that the liquid jet emerges at an angle of 45° to the horizontal, jumps over any low bund, returns to grade and forms a pool. It is then assumed that the pool spreads until the rate of spread is less than 1 mm/s and that the vapour from the edge of the pool then disperses as a Gaussian plume.

(b) Emission rate = 0.4 kg/s; length of liquid jet = 37 m; pool radius = 7 m; plume length to LEL = 11 m; plume length to 0.5LFL = 15 m; total distance to LFL = 55 m; total distance to 0.5LFL = 59 m.

(c) Emission rate = 0.6 kg/s; length of liquid jet = 84 m; pool radius = 9 m; plume length to LEL = 14 m; plume length to 0.5LFL = 19 m; total distance to LFL = 107 m; total distance to 0.5LFL = 113 m.

(d) Emission rate = 0.4 kg/s; length of liquid jet = 19 m; pool radius = 7 m; plume length to LEL = 11 m; plume length to 0.5LFL = 15 m; total distance to LFL = 37 m; total distance to 0.5LFL = 41 m.

(e) Emission rate = 0.6 kg/s; length of liquid jet = 37 m; pool radius = 9 m; plume length to LEL = 14 m; plume length to 0.5LFL = 19 m; total distance to LFL = 60 m; total distance to 0.5LFL = 66 m.

(f) For this case it is assumed that the liquid flashes off forming vapour and spray, that the latter also vaporises and that the resultant vapour disperses as a Gaussian plume.

with

$$t = \left(\frac{ku\sin\alpha}{kg} + g\right)[1 - \exp(-kt)] \tag{A12.3}$$

where k is a constant representing the air resistance.

The constant k has been estimated from the drag coefficient C_D, which has been taken as 0.3. This is a typical value used for plumes (eg Ooms, Mahieu and Zelis, 1974).

169

Liquid jets are also affected by jet breakup. Breakup generally starts quite a short distance from the exit, but no correlation has been found for the distance travelled by such a jet. There is, however, a rule of thumb that a practical jet, subject both to air resistance and breakup, travels only half the distance estimated neglecting both these factors.

The distance travelled by a flammable fluid before it falls below its LFL exceeds the throw of the jet. When the jet returns to grade the liquid forms a pool. Liquid is vaporised from the pool and the vapour disperses. The vaporisation has been calculated using the model for vaporisation of a volatile liquid and the dispersion using the Gaussian plume model.

For a flashing liquid jet it is assumed that the jet breaks up and that the droplets evaporate completely. The emission from the hole is calculated assuming that the fluid is a liquid, but the dispersion is calculated assuming that it vaporises instantaneously at the exit, using the Gaussian plume gas dispersion model given in equation (12.26).

The results are shown in Table A12.1.

Further details of the assumptions made and the models used are given in footnotes to the table.

NOTATION

C_D drag coefficient
g acceleration due to gravity
k constant representing air resistance
t time
u initial velocity of jet
x distance travelled by jet horizontal direction
z distance travelled by jet in vertical direction
α angle between jet and horizontal

APPENDIX 13—AN IGNITION MODEL

The estimates of the probability of ignition given in Section 15 are based on historical data and expert judgements of the probability of ignitions given the current mix of hazardous area classification practices.

It must be the ultimate aim to be able to predict the effect of making changes in these practices. It would greatly assist in making such predictions if the overall probability of ignition could be broken down into its constituent elements, or, in other words, if an ignition model could be developed.

Some features which such an ignition model should possess include the following:

1. The effect of the fluid phase, gas or liquid, should be taken into account.

2. The probability of ignition for an unprotected location should increase as the leak flow increases.

3. There should be a finite probability of ignition which allows for the failure to achieve perfect protection even within the designated zones and this probability also should increase as the leak flow increases.

4. There should be a finite probability of self-ignition.

5. The estimated probability of ignition should be the sum of (3) and (4) over the range covered by zoning and of (2) and (4) over the range not so covered.

6. The probability of ignition so estimated should be in broad agreement with that observed.

7. The contribution of self-ignition to the probability of ignition so estimated should be in broad agreement with that observed.

Self-ignition is used here to denote ignition at the point of release by whatever means. In other words, it denotes distance-independent ignition. Such ignition covers not only true autoignition, but cases of ignition by static electricity, by hot, distressed surfaces at the leak itself, by sparks associated with the leak rupture, etc. It does not, however, cover ignition by sources such as hot surfaces or electrical equipment, even those very close to the leak point.

Two separate ignition models have been produced, one for leaks of gas, or vapour, or of vapour and liquid, and one for leaks of liquid.

Figure A13.1 shows one attempt to construct an ignition model for gas or vapour. The figure is given with numerical values included but these are for illustrative purposes only. Line A is the crude observed ignition probability and is based on the line given in Figure 15.1 for gas leaks, which is simply a straight line drawn between the points for massive and minor leaks. Line B is the ignition probability due to the fact that the attempt to control ignition sources is not totally successful; it is thus termed the partial protection line. Line C is the ignition probability due to self-ignition. Line A is the sum of lines B and C. Curve D is an estimate of the ignition probability given no attempt to control ignition sources. All these are straight lines except that D curves over towards the right and approaches, but does not reach, a value close to unity. The actual ignition probability may not follow a straight line. Curve E illustrates a possible actual curve for observed ignition probability, which shows an increase in ignition probability at a point where the leak flow approaches values which the codes do not attempt to cater for.

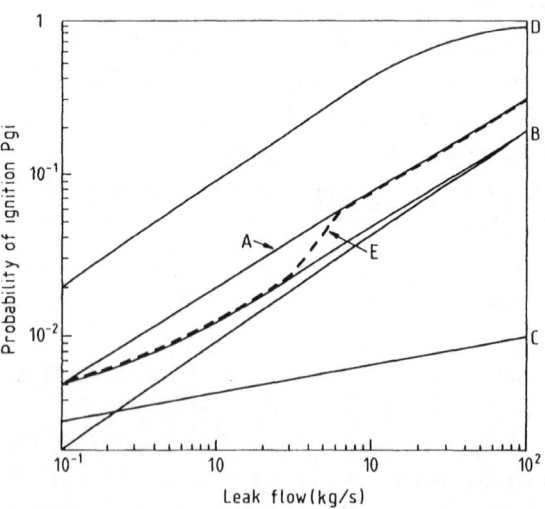

Figure A13.1   A speculative ignition model for leaks of gas

Attempts have been made to relate the slopes of the lines to the areas covered by releases of different sizes using relations such as those given by Marshall (1977, 1980) and assuming a constant density of ignition sources.

The attempt to construct such models brings home several important features of ignition probability. One is that unless the probability of self-ignition increases markedly with leak size, it must be an important contributor to small leaks. Another is that an important feature of the control of ignition sources is not simply that the distance between leak source and ignition source is increased, but the overall density of ignition sources is much reduced.

The models are therefore useful in highlighting features on which further work needs to be done, but they are considered too speculative to discuss here in any greater detail.

APPENDIX 14—STUDY OF FIRES AND VAPOUR CLOUD EXPLOSIONS

Although the prevention of a major vapour cloud explosion (VCE) is primarily a matter of preventing loss of containment and although hazardous area classification is not primarily concerned with events of this scale, nevertheless the VCE hazard must be considered in hazardous area classification.

In this appendix an attempt is made to estimate the frequency of major fires and vapour cloud explosions (VCEs) and to determine the distribution of leak sources for the latter.

A14.1 FIRES

Information on refinery fires is given in the annual series Reported Fire Losses in the Petroleum Industry issued by the API. Recent reports (API 1983, 1984 and 1985) give the number of fires in refining in the US in the years 1982–85 as 201, 173, 142 and 104, respectively.

The 1985 report gives the size of fires for 1985 as

Size of loss (1000$)	No. of fires	Proportion (%)
2.5–100	65	60
100–1000	37	34
> 1000	7	6

From Appendix 6 there were in the US in 1985 some 225 refineries. For the four-year period 1982–85 there were on average 155 fires/y. From the fire size distribution given above some 40% of fires may be regarded as large. Then the number of large fires averages 62 fires/y or 0.28 fires/ refinery y.

The Canvey Report (HSE, 1978) gives the frequency of a major refinery fire as

Frequency of major fire in refinery = 0.1 fires/refinery y

This is broadly consistent with the API data if only a proportion of the second category of fire is included.

A14.2 VAPOUR CLOUD EXPLOSIONS

There is no comparable statistical information on vapour cloud explosions, but individual surveys have been reported. Surveys of VCE incidents have been given by Davenport (1977) and (1983). Using only the second survey, the breakdown of incidents given by Davenport is as follows.

For vapour cloud explosions at fixed installations over the period 1962–82 inclusive Davenport (1983) records some 35 cases. The location of these explosions is

United States	19
Western Europe	9
Other	6

If the installations at risk in the US and Western Europe are taken as plants of the type of refinery units, as described in Section 9, the number at risk is 3182. Then

Frequency of vapour cloud explosions $= (19 + 9)/(21 \times 3182)$
$$= 4.2 \times 10^{-4} \text{ explosions/plant y}$$

From the table of incidents given in Davenport's paper the leak sources in these incidents may be obtained as shown in Table A14.1.

The incidents in Table A14.1 for which information is given may be broken down as shown in Table A14.2

Table A14.1 Leak sources in vapour cloud explosion incidents for 1962–1982 listed by Davenport (1983) – 1

Incident code no.	Leak source
76	Tank – reaction
138	Tank (refrigerated storage) – overpressure
1	Sight glass?
18	Pipe fitting
11	Pipe connection
7	Flare header
4	Pipe rupture
62	Pipe rupture
17	Valve – opened on air failure
6	Valve – bonnet failure
13	Tank – frothover
40	Pipe rupture
77	Reactor – reaction
32	Pipe connection
22	Pump
49	Flange
52	Valve – yoke failure
68	Hose rupture
9	Pipe rupture
57	Flange (on thermowell)
8	Expansion joint
118	Pipe?
60	Flare header rupture
70	Pipe rupture
74	Vessel – failure
51	Mixing nozzle
136	Pipe rupture
88	Tank (refrigerated storage) – failure
93	Pipe failure
94	Pipe failure
95	Pipe failure
81	Pipe elbow
99	Plug valve
141	Tank – overfilling

Table A14.2   Leak sources in vapour cloud explosion incidents for 1962–1982 listed by Davenport (1983) – 2

Leak source	No.	Proportion (%)
Tanks – reaction	1	2.9
Tanks – overfilling, frothover	2	5.9
Tanks – refrigerated storage, failure	2	5.9
Vessels – pressure vessel rupture	1	2.9
Reactor – reaction	1	2.9
Pipe	10	29.4
Flange	2	5.9
Other fittings	6	29.4
Hose	1	2.9
Valve	3	8.8
Sight glass	1	2.9
Pump	1	2.9
Valve opened	1	2.9
Flare	2	5.9
Total	34	100.0

APPENDIX 15—FIRE AND EXPLOSION MODEL NUMERICAL INVESTIGATION

A numerical investigation has been carried out to determine whether the leak frequency and leak size values given in Section 11 and Appendix 8 and the ignition and explosion probabilities given in Sections 15 and 16 are consistent with the fire and explosion estimates given in Section 17 and Appendix 14.

The basis of the investigation is the profile of the equivalent standard plant given in Table 8.2.

The leak frequencies and leak sizes used for pipes, flanges and valves used are those given in Table 11.1.

For rupture of pipework the hole size was taken as 1.5 times the cross-sectional area of the pipe. This allows for a mix of situations, some with sufficient inventory in both sections of the pipe to sustain discharge and others with sufficient inventory in one section of the pipe only.

The leak flows for liquids were determined from equation (12.5) using the following data:

Density of liquid $= 600 \text{ kg/m}^3$
Pressure in plant $= 15 \text{ bar a.}$
Coefficient of discharge $C_D = 0.8$ for pipe
$= 0.6$ for hole

Those for gases were determined from equation (12.1) using the following data:

Molecular weight of gas $= 44$
Pressure in plant $= 15 \text{ bar a.}$
Expansion index $= 1.4$
Coefficient of discharge $C_D = 0.8$ for pipe
$= 0.6$ for hole

A cutoff was applied to the ignition probabilities, the lowest value used for gas and liquids being 0.01, the highest value for gas being 0.3, and the highest value for liquids 0.08. Similarly for the probability of explosion given ignition for a gas the highest value used was 0.25.

Then for each leak source – pipe, flange, valve, pump and small bore connection – for each leak size the following quantities were determined:

Leak frequency per unit (per m, per flange, per valve)

Leak frequency per plant

Mass flow

Ignition probability

Explosion probability (gases, flashing liquids)

Fire frequency

Explosion frequency (gases, flashing liquids)

The principal parameters used are summarised in Table 18.1 and the leak areas are given in Table A15.1. The results are shown by individual fluid phase – liquid, gas or flashing liquid – in Tables A15.3 – A15.4 and summaries of the results by leak source and by fluid phase are given in Tables 18.2 and 18.3, respectively.

These results are discussed in Section 18.

Table A15.1   Fire and explosion model base case: leak size distributions

All figures to be multiplied by 10^{-3}

A Pipework

Pipe diameter (m)	Leak area (m^2)			
	0.025	0.050	0.100	0.300
Rupture leak	0.491	1.96	7.85	70.7
Major leak	0.0491	0.196	0.785	7.07
Minor leak	0.0049	0.0196	0.0785	0.707

B Flanges

Pipe diameter (m)	Leak area (m^2)			
	0.025	0.050	0.100	0.300
Major leak	0.0785	0.0785	0.105	0.189
Minor leak	0.0079	0.0079	0.0105	0.0188

C Valves

	Leak area (m^2)			
Pipe diameter (m)	0.025	0.050	0.100	0.300
Rupture leak	0.491	1.96	7.85	70.7
Major leak	0.0491	0.196	0.785	7.07
Minor leak	0.0049	0.0196	0.0785	0.707

D Pumps

	Leak area (m^2)			
Pipe diameter (m)	0.025	0.050	0.100	0.300
Rupture leak	0.491	1.96	7.85	70.7
Major leak	0.0491	0.196	0.785	7.07
Minor leak	0.0049	0.0196	0.0785	0.707

E Small bore connections

	Leak area (m^2)
Pipe diameter (m)	0.010
Rupture leak	0.079
Major leak	0.0079

Table A15.2   Fire and explosion model base case: liquid leaks

A Pipework

Pipe diameter (m)	0.025	0.050	0.100	0.300
Pipe lengths (m)	1875.0	1680.0	2160.0	577.5

	Leak frequency (leaks/m y $\times$ 10^4)			
Pipe diameter (m)	0.025	0.050	0.100	0.300
Rupture leak	0.005000	0.00500	0.00150	0.00050
Major leak	0.050000	0.05000	0.01500	0.00500
Minor leak	0.50000	0.50000	0.15000	0.05000

	Leak frequency (leaks/plant y)			
Pipe diameter (m)	0.025	0.050	0.100	0.300
Rupture leak	0.00094	0.00084	0.00032	0.00003
Major leak	0.00938	0.00840	0.00324	0.00029
Minor leak	0.09375	0.08400	0.03240	0.00289

Mass flow (at 15 bar a) (kg/s)

Pipe diameter (m)	0.025	0.050	0.100	0.300
Rupture leak	24.14	96.57	386.30	3476.70
Major leak	1.21	4.83	19.32	173.84
Minor leak	0.12	0.48	1.93	17.38

Ignition probability

Pipe diameter (m)	0.025	0.050	0.100	0.300
Rupture leak	0.04580	0.07891	0.08000	0.08000
Major leak	0.01413	0.02435	0.04196	0.08000
Minor leak	0.01000	0.01000	0.01700	0.04026

Fire frequency (fires/plants $y \times 10^4$)

Pipe diameter (m)	0.025	0.050	0.100	0.300
Rupture leak	0.429	0.663	0.259	0.023
Major leak	1.325	2.046	1.359	0.231
Minor leak	9.375	8.400	5.507	1.162

B Flanges

Pipe diameter (m)	0.025	0.050	0.100	0.300
Number	450	428	324	77

Leak frequency (leaks/flange $y \times 10^4$)

Pipe diameter (m)	0.025	0.050	0.100	0.300
Major leak	0.3000	0.3000	0.3000	0.3000
Minor leak	3.0000	3.0000	3.0000	3.0000

Leak frequency (leaks/plant y)

Pipe diameter (m)	0.025	0.050	0.100	0.300
Major leak	0.01350	0.01284	0.00972	0.00231
Minor leak	0.13500	0.12840	0.09720	0.02310

Mass flow (15 bar a) (kg/s)

Pipe diameter (m)	0.025	0.050	0.100	0.300
Major leak	1.93	1.93	2.58	4.64
Minor leak	0.19	0.19	0.26	0.46

Ignition probability

Pipe diameter (m)	0.025	0.050	0.100	0.300
Major leak	0.01700	0.01700	0.01903	0.02396
Minor leak	0.01000	0.01000	0.01000	0.01000

Fire frequency (fires/plant y × 10⁴)

Pipe diameter (m)	0.025	0.050	0.100	0.300
Major leak	2.294	2.182	1.849	0.554
Minor leak	13.500	12.840	9.720	2.310

C Valves

Pipe diameter (m)	0.025	0.050	0.100	0.300
Number	360	200	96	14

Leak frequency (leaks/valve y × 10⁴)

Pipe diameter (m)	0.025	0.050	0.100	0.300
Rupture leak	0.0100	0.0100	0.0100	0.0050
Major leak	0.1000	0.1000	0.1000	0.0500
Minor leak	1.0000	1.0000	1.0000	0.5000

Leak frequency (leaks/plant y)

Pipe diameter (m)	0.025	0.050	0.100	0.300
Rupture leak	0.00036	0.00020	0.00010	0.00001
Major leak	0.00360	0.00200	0.00096	0.00007
Minor leak	0.03600	0.02000	0.00960	0.00070

Mass flow (kg/s)

Pipe diameter (m)	0.025	0.050	0.100	0.300
Rupture leak	12.07	48.29	193.15	1738.35
Major leak	1.21	4.83	19.32	173.84
Minor leak	0.12	0.48	1.93	17.38

Ignition probability

Pipe diameter (m)	0.025	0.050	0.100	0.300
Rupture leak	0.03489	0.06012	0.08000	0.08000
Major leak	0.01413	0.02435	0.04196	0.08000
Minor leak	0.01000	0.01000	0.01700	0.04026

Fire frequency (fires/plant y × 10⁴)

Pipe diameter (m)	0.025	0.050	0.100	0.300
Rupture leak	0.126	0.120	0.077	0.006
Major leak	0.509	0.487	0.403	0.056
Minor leak	3.600	2.000	1.632	0.282

D Pumps

Pipe diameter (m)	0.025	0.050	0.100	0.300
Number	0	11	6	0

Leak frequency (leak/pump y × 10^4)

Pipe diameter (m)	0.025	0.050	0.100	0.300
Rupture leak	0.3000	0.3000	0.3000	0.3000
Major leak	3.0000	3.0000	3.0000	3.0000
Minor leak	30.0000	30.0000	30.0000	30.0000

Leak frequency (leaks/plant y)

Pipe diameter (m)	0.025	0.050	0.100	0.300
Rupture leak	0.00000	0.00033	0.00018	0.00000
Major leak	0.00000	0.00330	0.00180	0.00000
Minor leak	0.00000	0.03300	0.01800	0.00000

Mass flow (kg/s)

Pipe diameter (m)	0.025	0.050	0.100	0.300
Rupture leak	12.07	48.29	193.15	1738.35
Major leak	1.21	4.83	19.32	173.84
Minor leak	0.12	0.48	1.93	17.38

Ignition probability

Pipe diameter (m)	0.025	0.050	0.100	0.300
Rupture leak	0.03489	0.06012	0.08000	0.08000
Major leak	0.01413	0.02435	0.04196	0.08000
Minor leak	0.01000	0.01000	0.01700	0.04026

Fire frequency (fires/plant y × 10^4)

Pipe diameter (m)	0.025	0.050	0.100	0.300
Rupture leak	0.000	0.198	0.144	0.000
Major leak	0.000	0.804	0.755	0.000
Minor leak	0.000	3.300	3.059	0.000

E Small bore connections

Pipe diameter (m)	0.010
Number	280

Leak frequency (leaks/connection y × 10^4)

Pipe diameter (m)	0.010
Rupture leak	1.0
Major leak	10.0

Leak frequency (leaks/plant y)

Pipe diameter (m)	0.010
Rupture leak	0.028
Major leak	0.28

Mass flow (kg/s)

Pipe diameter (m)	0.010
Rupture leak	2.5753
Major leak	0.1932

Ignition probability

Pipe diameter (m)	0.010
Rupture leak	0.01903
Major leak	0.01000

Fire frequency (fires/plant y $\times$ 10^4)

Pipe diameter (m)	0.010
Rupture leak	0.001
Major leak	0.003

Table A15.3   Fire and explosion model base case: gas leaks

A Pipework

Pipe diameter (m)	0.025	0.050	0.100	0.100
Pipe length (m)	0	630	810	413

Leak frequency (leaks/plant y)

Pipe diameter (m)	0.025	0.050	0.100	0.300
Rupture leak	0.00000	0.00031	0.00012	0.00002
Major leak	0.00000	0.00315	0.00121	0.00021
Minor leak	0.00000	0.03150	0.01215	0.00206

Mass flow (kg/s)

Pipe diameter (m)	0.025	0.050	0.100	0.300
Rupture leak	2.543	10.173	40.694	366.242
Major leak	0.127	0.509	2.035	18.312
Minor leak	0.013	0.051	0.203	1.831

Ignition probability

Pipe diameter (m)	0.025	0.050	0.100	0.300
Rupture leak	0.02841	0.06918	0.16845	0.30000
Major leak	0.01000	0.01011	0.02462	0.10089
Minor leak	0.01000	0.01000	0.01000	0.02301

Fire frequency (fires/plant y $\times$ 10^4)

Pipe diameter (m)	0.025	0.050	0.100	0.300
Rupture leak	0.000	0.218	0.205	0.062
Major leak	0.000	0.318	0.299	0.208
Minor leak	0.000	3.150	1.215	0.475

Explosion probability

Pipe diameter (m)	0.025	0.050	0.100	0.300
Rupture leak	0.00144	0.00641	0.02849	0.07500
Major leak	0.00014	0.00025	0.00113	0.01206
Minor leak	0.00005	0.00009	0.00017	0.00101

Explosion frequency (explosions/plant $y \times 10^4$)

Pipe diameter (m)	0.025	0.050	0.100	0.300
Rupture leak	0.000	0.020	0.035	0.015
Major leak	0.000	0.008	0.014	0.025
Minor leak	0.000	0.029	0.021	0.021

B Flanges

Pipe diameter (m)	0.025	0.050	0.100	0.300
Number	0	160	122	50

Leak frequency (leaks/plant y)

Pipe diameter (m)	0.025	0.050	0.100	0.300
Major leak	0.00000	0.00481	0.00364	0.00165
Minor leak	0.00000	0.04815	0.03645	0.01650

Mass flow (kg/s)

Pipe diameter (m)	0.025	0.050	0.100	0.300
Major leak	0.203	0.203	0.271	0.488
Minor leak	0.020	0.020	0.027	0.049

Ignition probability

Pipe diameter (m)	0.025	0.050	0.100	0.300
Major leak	0.01000	0.01000	0.01000	0.01000
Minor leak	0.01000	0.01000	0.01000	0.01000

Fire frequency (fires/plant $y \times 10^4$)

Pipe diameter (m)	0.025	0.050	0.100	0.300
Major leak	0.000	0.480	0.365	0.165
Minor leak	0.000	4.815	3.645	1.650

Explosion probability

Pipe diameter (m)	0.025	0.050	0.100	0.300
Major leak	0.00017	0.00017	0.00019	0.00025
Minor leak	0.00006	0.00006	0.00007	0.00009

Explosion frequency (explosions/plant $y \times 10^4$)

Pipe diameter (m)	0.025	0.050	0.100	0.300
Major leak	0.000	0.008	0.007	0.004
Minor leak	0.000	0.030	0.026	0.015

C Valves

Pipe diameter (m)	0.025	0.050	0.100	0.300
Number	0	75	36	10

Leak frequency (leaks/plant y)

Pipe diameter (m)	0.025	0.050	0.100	0.300
Rupture leak	0.00000	0.00008	0.00004	0.00000
Major leak	0.00000	0.00075	0.00036	0.00005
Minor leak	0.00000	0.00750	0.00360	0.00050

Mass flow (kg/s)

Pipe diameter (m)	0.025	0.050	0.100	0.300
Rupture leak	1.272	5.087	20.347	183.122
Major leak	0.127	0.509	2.035	18.312
Minor leak	0.013	0.051	0.203	1.831

Ignition probability

Pipe diameter (m)	0.025	0.050	0.100	0.300
Rupture leak	0.01821	0.04433	0.10795	0.30000
Major leak	0.01000	0.01011	0.02462	0.10089
Minor leak	0.01000	0.01000	0.01000	0.02301

Fire frequency (fires/plant y $\times$ 10^4)

Pipe diameter (m)	0.025	0.050	0.100	0.300
Rupture leak	0.000	0.033	0.039	0.015
Major leak	0.000	0.076	0.089	0.050
Minor leak	0.000	0.750	0.360	0.115

Explosion probability

Pipe diameter (m)	0.025	0.050	0.100	0.300
Rupture leak	0.00068	0.00304	0.01351	0.07500
Major leak	0.00014	0.00025	0.00113	0.01206
Minor leak	0.00005	0.00009	0.00017	0.00101

Explosion frequency (explosions/plant y $\times$ 10^4)

Pipe diameter (m)	0.025	0.050	0.100	0.300
Rupture leak	0.000	0.002	0.005	0.004
Major leak	0.000	0.002	0.004	0.006
Minor leak	0.000	0.007	0.006	0.005

D Small bore connections

Pipe diameter (m)	*0.010*
Number	105

Leak frequency (leaks/plant y)

Pipe diameter (m)	*0.010*
Rupture leak	0.01050
Major leak	0.10500

Mass flow (kg/s)

Pipe diameter (m)	*0.010*
Rupture leak	0.2713
Major leak	0.0203

Ignition probability

Pipe diameter (m)	*0.010*
Rupture leak	0.01000
Major leak	0.01000

Fire frequency (fires/plant y $\times$ 10^4)

Pipe diameter (m)	*0.010*
Rupture leak	1.050
Major leak	10.500

Explosion probability

Pipe diameter (m)	*0.010*
Rupture leak	0.00019
Major leak	0.00006

Explosion frequency (explosions/plant y $\times$ 10^4)

Pipe diameter (m)	*0.010*
Rupture leak	0.020
Major leak	0.065

Table A15.4 Fire and explosion model base case: two-phase leaks

A Pipework

Pipe diameter (m)	0.025	0.050	0.100	0.300
Pipe length (m)	1875.0	1890.0	2430.0	660.0

Leak frequency (leaks/plant y)

Pipe diameter (m)	0.025	0.050	0.100	0.300
Rupture leak	0.00094	0.00094	0.00036	0.00003
Major leak	0.00938	0.00945	0.00364	0.00330
Minor leak	0.09375	0.09450	0.03645	0.00330

Mass flow (kg/s)

Pipe diameter (m)	0.025	0.050	0.100	0.300
Rupture leak	19.31	77.26	309.04	2781.36
Major leak	0.97	3.86	15.45	139.07
Minor leak	0.10	0.39	1.55	13.91

Ignition probability

Pipe diameter (m)	0.025	0.050	0.100	0.300
Rupture leak	0.10440	0.25421	0.30000	0.30000
Major leak	0.01526	0.03715	0.09047	0.30000
Minor leak	0.01000	0.01000	0.02063	0.08455

Fire frequency (fires/plant y $\times$ 10^4)

Pipe diameter (m)	0.025	0.050	0.100	0.300
Rupture leak	0.979	2.402	1.094	0.099
Major leak	1.431	3.511	3.298	0.990
Minor leak	9.375	9.450	7.521	2.790

Explosion probability

Pipe diameter (m)	0.025	0.050	0.100	0.300
Rupture leak	0.01277	0.05681	0.07500	0.07500
Major leak	0.00051	0.00226	0.01005	0.07500
Minor leak	0.00012	0.00022	0.00084	0.00897

Explosion frequency (explosions/plant y $\times$ 10^4)

Pipe diameter (m)	0.025	0.050	0.100	0.300
Rupture leak	0.120	0.537	0.273	0.025
Major leak	0.048	0.213	0.366	0.247
Minor leak	0.115	0.211	0.307	0.296

B Flanges

Pipe diameter (m)	0.025	0.050	0.100	0.300
Number	450	482	364	83

Leak frequency (leaks/plant y)

Pipe diameter (m)	0.025	0.050	0.100	0.300
Major leak	0.01350	0.01444	0.01095	0.00264
Minor leak	0.13500	0.14445	0.10935	0.02640

Mass flow (kg/s)

Pipe diameter (m)	0.025	0.050	0.100	0.300
Major leak	1.545	1.545	2.060	3.708
Minor leak	0.155	0.155	0.206	0.371

Ignition probability

Pipe diameter (m)	0.025	0.050	0.100	0.300
Major leak	0.02063	0.02063	0.02482	0.03619
Minor leak	0.01000	0.01000	0.01000	0.01000

Fire frequency (fires/plant $y \times 10^4$)

Pipe diameter (m)	0.025	0.050	0.100	0.300
Major leak	2.785	2.980	2.714	0.956
Minor leak	13.500	14.445	10.935	2.640

Explosion probability

Pipe diameter (m)	0.025	0.050	0.100	0.300
Major leak	0.00084	0.00084	0.00115	0.00216
Minor leak	0.00015	0.00015	0.00017	0.00022

Explosion frequency (explosions/plant $y \times 10^4$)

Pipe diameter (m)	0.025	0.050	0.100	0.300
Major leak	0.112	0.122	0.126	0.057
Minor leak	0.203	0.217	0.186	0.058

C Valves

Pipe diameter (m)	0.025	0.050	0.100	0.300
Number	360	225	108	16

Leak frequency (leaks/plant y)

Pipe diameter (m)	0.025	0.050	0.100	0.300
Rupture leak	0.00036	0.00022	0.00011	0.00001
Major leak	0.00360	0.00225	0.00108	0.00008
Minor leak	0.03600	0.02250	0.01080	0.00080

	Mass flow (kg/s)			
Pipe diameter (m)	0.025	0.050	0.100	0.300
Rupture leak	9.66	38.63	154.52	1390.68
Major leak	0.97	3.86	15.45	139.07
Minor leak	0.10	0.39	1.55	13.91

	Ignition probability			
Pipe diameter (m)	0.025	0.050	0.100	0.300
Rupture leak	0.06691	0.16291	0.30000	0.30000
Major leak	0.01526	0.03715	0.09047	0.30000
Minor leak	0.01000	0.01000	0.02063	0.08455

	Fire frequency (fires/plant y $\times 10^4$)			
Pipe diameter (m)	0.025	0.050	0.100	0.300
Rupture leak	0.241	0.367	0.324	0.024
Major leak	0.549	0.836	0.977	0.240
Minor leak	3.600	2.250	2.228	0.676

	Explosion probability			
Pipe diameter (m)	0.025	0.050	0.100	0.300
Rupture leak	0.00606	0.02694	0.07500	0.07500
Major leak	0.00051	0.00226	0.01005	0.07500
Minor leak	0.00012	0.00022	0.00084	0.00897

	Explosion frequency (explosions/plant y $\times 10^4$)			
Pipe diameter (m)	0.025	0.050	0.100	0.300
Rupture leak	0.022	0.061	0.081	0.006
Major leak	0.018	0.051	0.108	0.060
Minor leak	0.044	0.050	0.091	0.072

D Pumps

Pipe diameter (m)	0.025	0.050	0.100	0.300
Number	0	4	4	0

	Leak frequency (leaks/plant y)			
Pipe diameter (m)	0.025	0.050	0.100	0.300
Rupture leak	0.00000	0.00012	0.00012	0.00000
Major leak	0.00000	0.00120	0.00120	0.00000
Minor leak	0.00000	0.01200	0.01200	0.00000

	Mass flow (kg/s)			
Pipe diameter (m)	0.025	0.050	0.100	0.300
Rupture leak	9.66	38.63	154.52	1390.68
Major leak	0.97	3.86	15.45	139.07
Minor leak	0.10	0.39	1.55	13.91

Ignition probability

Pipe diameter (m)	0.025	0.050	0.100	0.300
Rupture leak	0.06691	0.16291	0.30000	0.30000
Major leak	0.01526	0.03715	0.09047	0.30000
Minor leak	0.01000	0.01000	0.02063	0.08455

Fire frequency (fires/plant $y \times 10^4$)

Pipe diameter (m)	0.025	0.050	0.100	0.300
Rupture leak	0.000	0.195	0.360	0.000
Major leak	0.000	0.446	1.086	0.000
Minor leak	0.000	1.200	2.476	0.000

Explosion probability

Pipe diameter (m)	0.025	0.050	0.100	0.300
Rupture leak	0.00606	0.02694	0.07500	0.07500
Major leak	0.00051	0.00226	0.01005	0.07500
Minor leak	0.00012	0.00022	0.00084	0.00897

Explosion frequency (explosions/plant $y \times 10^4$)

Pipe diameter (m)	0.025	0.050	0.100	0.300
Rupture leak	0.000	0.032	0.090	0.000
Major leak	0.000	0.027	0.121	0.000
Minor leak	0.000	0.027	0.101	0.000

E Small bore connections

Pipe diameter (m)	0.010
Number	315

Leak frequency (leaks/plant y)

Pipe diameter (m)	0.010
Rupture leak	0.03150
Major leak	0.31500

Mass flow (kg/s)

Pipe diameter (m)	0.010
Rupture leak	2.0603
Major leak	0.1545

Ignition probability

Pipe diameter (m)	0.010
Rupture leak	0.02482
Major leak	0.01000

Fire frequency (fires/plant y $\times$ 10^4)

Pipe diameter (m)	*0.010*
Rupture leak	7.818
Major leak	31.500

Explosion probability

Pipe diameter (m)	*0.010*
Rupture leak	0.00115
Major leak	0.00015

Explosion frequency (explosions/plant y $\times$ 10^4)

Pipe diameter (m)	*0.010*
Rupture leak	0.020
Major leak	0.065

REFERENCES

Anon. (1985). *Oil Gas J.*, 83 Dec.30, 100.

Anyakora, S.N., Engel, G.F.M. and Lees, F.P. (1971). Some data on the reliability of instruments in the chemical plant environment. *Chem. Engr*, 255, 396.

American Petroleum Institute (1984). *Reported fire losses in the petroleum industry for 1983* (Washington, DC).

American Petroleum Institute (1985). *Reported fire losses in the petroleum industry for 1984* (Washington, DC).

American Petroleum Institute (1986). *Reported fire losses in the petroleum industry for 1985* (Washington, DC).

Arulanantham, D.C. and Lees, F.P. (1981). Some data on the reliability of pressure equipment in the chemical plant environment. *Int J. Pres. Ves. Piping*, 9, 327.

Aupied, J.R., Le Coguiec, A. and Procaccia, H. (1983). Valves and pumps operating experience in French nuclear plants. *Reliab. Engng*, 6, 133.

Batstone, R.J. and Tomi, D.J. (1980). Hazard analysis in planning industrial developments. *Loss Prevention*, vol.13 (New York: Am. Inst. Chem. Engrs).

Beattie, J.R. (1963). *An assessment of environmental hazards from fission product releases*. UKAEA, Health and Safety Branch, Rep. AHSB(S) R64 (Culcheth, Warrington, Lancashire).

Bochinski, J.H., Schoultz, K.S.and Gideon, J.A. (1979). Meeting emission standards for acrylonitrile. *Chem. Engng Prog.*, 75(8), 53.

Bondi, A. (1966). Estimation of the heat capacity of liquids. Ind. Engng *Chem. Fundam.*, 5, 443.

Brown, B. (1969). *General properties of matter* (London: Butterworths)

Browning, R.L. (1969). Estimating loss probabilities. *Chem. Engng*, 76, Dec.15, 135.

Browning, R.L. (1980). *The loss rate concept in safety engineering* (New York: Dekker).

Bush, S.H. (1977). A review of reliability of piping in light-water reactors. In *Failure data and failure analysis in power and processing industries* (edited by A.C.Gangadharan and S.J.Brown) (New York: Am. Soc. Mech. Engrs), p.37.

Chen, C.J. and Rodi, W. (1980). *Vertical turbulent buoyant jets* (Oxford: Pergamon).

Chemical Rubber Co. (1979). *CRC Handbook of chemistry and physics*, 60th ed. (Chemical Rubber Co.).

Chung, T.H., Lee, L.L. and Starling, K.E. (1984). Applications of kinetic gas theories and multiparameter correlation for prediction of dilute gas viscosity and thermal conductivity. *Ind. Engng Chem. Fundam.*, 23, 8.

Dahl, E. (1983). *Risk of oil and gas blowout on the Norwegian Continental Shelf.* SINTEF, Trondheim, Rep.STF 88A82062.

Davenport, J. (1977). A survey of vapor cloud incidents. In *Loss Prevention*, vol.11 (New York: Am. Inst. Chem. Engrs), p.39.

Davenport, J.A. (1983). A study of vapor cloud incidents – an update. In *Loss Prevention and Safety Promotion*, vol.4, pt 1 (Rugby: Instn Chem. Engrs), p.C1.

Engh, T.A. and Larsen, K. (1979). Breakup length of turbulent water and steel jets in air. *Scand. J. Metall.*, no.8, 161.

Ewan, B.C.R. and Moodie, K. (1986). Structure and velocity measurements in underexpanded jets. *Combust. Sci. Technol.*, 45, 275.

Fauske, H.K. (1965). The discharge of saturated water through tubes. *Chem. Engng Prog. Symp. Ser.59* (New York: Am. Inst. Chem. Engrs), p.210.

Fauske, H.K. (1985). Flashing flows or some practical guidelines for emergency releases. *Plant/Operations Prog.*, 4, 132.

Fire Protection Association (1974). Increasing number and cost of fires in chemical process industries. *Fire Prevention*, 106, 15.

Fletcher, B. (1984). Flashing flow through orifices and pipes. *Chem. Engng Prog.*, 80(3), 76.

Fletcher, B. and Johnson, A.E. (1984). The discharge of superheated liquids from pipes. *The protection of exothermic reactors and pressurized storage vessels* (Rugby: Instn Chem. Engrs), p.149.

Forsth, L.R. (1981a). *Fires and explosions on fixed platforms in the Norwegian North Sea*. Det Norske Veritas, Rep.81–1175.

Forsth, L.R. (1981b). *Fires and explosions on offshore platforms in the Gulf of Mexico*. Det Norske Veritas, Rep.81–1221.

Forsth, L.R. (1983). Survey identifies causes of offshore fires and explosions. *Oil Gas. J.*, 81 July 25, 154.

Freeberg, C.R. and Aarni, C.W. (1982). Hydrocarbon emission control in petroleum refineries. *Chem. Engng Prog.*, 78(6), 35.

Gifford, F.A. (1961). Use of routine meteorological observations for estimating atmospheric dispersion. *Nucl. Safety*, 2(4), 47.

Gugan, K. (1978). *Unconfined vapour cloud explosions* (London: Instn Chem. Engrs and Godwin).

Hansen, J., de Heer, H.J. and Kortlandt, D. (1980). Comparative risk analysis of processing plant. *Loss Prevention and Safety Promotion in the Process Industries 3* vol.2, p.455 (Basle: Swiss Society of Chemical Industries).

Harris, R.J. (1983). *Gas explosions in buildings and heating plant* (London: Spon).

Hawksley, J.L. (1984). Some social, technical and economical aspects of the risks of large chemical plants. *CHEMRAWN III*.

Health and Safety Executive (1978). *Canvey: An investigation of potential hazards from operations in the Canvey Island/Thurrock area* (London: HM Stationery Office).

Health and Safety Executive (1981). *Canvey. A second report: A review of potential hazards from operations in the Canvey Island/Thurrock area three years after publication of the Canvey Report* (London: HM Stationery Office).

Health and Safety Executive (1987). *Health and safety statistics 1985–86* (London: HM Stationery Office).

Health and Safety Executive (1988). *The tolerability of risk from nuclear power stations* (London: HM Stationery Office).

Health and Safety Executive (1989). *Risk criteria for land-use planning in the vicinity of major industrial hazards* (London: HM Stationery Office).

HM Chief Inspector of Factories (1986). *Health and Safety at Work. Report of HM Chief Inspector of Factories 1985* (London: HM Stationery Office).

Hooper, W.B. (1982). Predict fittings for piping systems. *Chem. Engng*, 89, May17, 127.

Van der Hoven, I. (1976). A survey of field measurements of atmospheric diffusion under low wind speed, inversion conditions. *Nucl. Safety*, 17, 223.

Hughes, T.W., Tierney, D.R. and Khan, Z.S. (1979). Measuring fugitive emissions from petrochemical plants. *Chem. Engng Prog.*, 75(8), 35.

Institution of Electrical Engineers (1971). *Electrical safety in hazardous environments 1* (London).

Institution of Electrical Engineers (1975). *Electrical safety in hazardous environments 2* (London).

Institution of Electrical Engineers (1982a). *Electrical safety in hazardous environments 3* (London).

Institution of Electrical Engineers (1982b). *Flammable atmospheres and area classification*, Colloquium Digest 1982/26 (London).

Institution of Electrical Engineers (1988). *Electrical safety in hazardous environments 4* (London).

Jones, A.L. (1984). Fugitive emissions of volatile hydrocarbons. *Chem. Engr*, 406 Aug/Sept., 12.

Jones, M.R.O. and Underwood, M.C. (1983). An appraisal of expressions used to calculate the release rate of pressurised liquefied gases. *Chem. Engng J.*, 26, 251.

Katan, L.L. (1951). *The fire hazard of fuelling aircraft in the open.* Fire Res. Tech. Paper 1 (London: HM Stationery Office).

Kletz, T.A. (1971). Hazard analysis – a quantitative approach to safety. In *Major loss prevention in the process industries* (London: Instn Chem. Engrs), p.75.

Kletz, T.A. (1984). Are safety valves old hat? *Chem. Processing* 20(9), 77.

Kletz, T.A. (1977). Unconfined vapour cloud explosions. In *Loss Prevention*, vol.11 (New York: Am. Inst. Chem. Engrs), p.50.

Kletz, T.A. (1984). The prevention of major leaks – better inspection after construction? *Plant/Operations Prog.*, 3, 19.

Kunstmann, G. (1980). Leak rates at valve stems. *Verfahrenstechnik*, 14, 260.

Lawley, H.G. and Kletz, T.A. (1975). High-pressure trip systems for vessel protection. *Chem. Engng*, 82 May12, 81.

Lees, F.P. (1980). *Loss prevention in the process industries* (London: Butterworths).

Lipton, S. and Lynch, J. (1987). *Health hazard control in the chemical process industry* (New York: Wiley).

Long, V.D. (1963). Estimation of the extent of hazard areas around a vent. *Hazards 3* (London: Instn Chem. Engrs), p.6.

Mackay, D. and Matsugu, R.S. (1973). Evaporation rates of liquid hydrocarbon spills on land and water. *Can. J. Chem. Engng*, 51, 434.

McMullen, R.W. (1975). The change of concentration standard deviations with distance. *J. Air Poll. Control Ass.*, 25, 1057.

Marshall, J.G. (1977). Hazardous clouds and flames arising from continuous releases into the atmosphere. *Chemical process hazards 6* (Rugby: Instn Chem. Engrs), p.21.

Marshall, J.G. (1980). The size of flammable clouds arising from continuous releases into the atmosphere – Part 2. *Chemical process hazards 7* (Rugby: Instn Chem. Engrs), p.11.

Mecklenburgh, J.C. (1985). *Process plant layout* (London: Godwin).

Morgester, J.J., Frisk, D.L., Zimmerman, G.L., Vincent, R.C. and Jordan, G.H. (1979). Control of emissions from refinery valves and flanges. *Chem. Engng Prog.*, 75(8), 40.

Nixon, J. (1971). 'Risk point' determination as an alternative to hazardous area classification: Part 2. In IEE (1971), *op.cit.*, p.150.

Ooms, G., Mahieu, A.P. and Zelis, F. (1974). The plume path of vent gases heavier than air. In *Loss Prevention and Safety Promotion in the Process Industries 1* (Amsterdam: Elsevier), p.211.

O'Shea, P.C. (1982). Flammable gases and liquids – a discussion of the parameters affecting their release and dispersion. In IEE (1982b), *op.cit.*

Palles-Clark, P.C. (1982). History of area classification and the basic content of BS 5345: Part 2. In IEE (1982b), *op.cit.*

Pape, R.P. and Nussey, C. (1985). A basic approach for the analysis of risks from major toxic hazards. *The assessment and control of major hazards* (Rugby: Instn Chem. Engrs), p.367.

Pape, R.P. (1989). In *Safety cases* (edited by F.P.Lees and M.L.Ang) (London: Butterworths).

Pasquill, F. (1943). Evaporation from a plane free-liquid surface into a turbulent air stream. *Proc. Roy. Soc.*, A182, 75.

Pasquill, F. (1961). The estimation of the dispersion of windborne materials. *Met. Mag.*, 90(1063), 33.

Pasquill, F. (1962). *Atmospheric diffusion* (London: van Nostrand).

Powell, R.W. (1984). Estimating worker exposure to gases and vapours leaking from pumps and valves. *Am. Ind. Hyg. Ass. J.*, 45, 47.

Rackett, H.G. (1970). Equations of state for saturated liquids. *J. Chem. Engng Data*, 15, 514.

Rajaratnam, N. (1976). *Turbulent jets* (Amsterdam: Elsevier).

Redpath, P.G. (1976). Fire: occurrence and effects. *Loughborough Univ. of Technol., Dept of Chem. Engng, Course on Loss Prevention in the Process Industries* (quoted in Lees, F.P. (1980)), *op.cit.*, Table 2.11.

Reid, R.C., Prausnitz, J.M. and Poling, B.E. (1987). *The properties of liquids and gases*, 4th ed. (New York: McGraw-Hill).

Ricou, F.P. and Spalding, D.B. (1961). Measurements of entrainment by axisymmetrical turbulent jets. *J. Fluid Mech.*, 11, 21.

Rijnmond Public Authority (1982). *Risk analysis of six potentially hazardous industrial objects in the Rijnmond area. A pilot study* (Dordrecht: Reidel).

Rosebrook, D.D. (1977). Fugitive hydrocarbon emissions. *Chem. Engng*, 84 Oct.17, 143.

Rowlinson, J.S. (1969). *Liquids and liquid mixtures*, 2nd ed. (London: Butterworths).

Sagendorf, J. (1975). *Diffusion under low wind speed and inversion conditions*. NOAA, Environ. Res. Labs, ERL-ARL Tech. Memo. 52.

Shaw, P. and Briscoe, F. (1978). *Evaporation from spills of hazardous liquids on land and water*. UKAEA, Safety and Reliability Directorate, Rep. SRD R 100 (Culcheth, Warrington, Lancs).

Sherwin, D.J. and Lees, F.P. (1980). An investigation of the application of failure data analysis to decision-making in maintenance of process plants. Pts 1–2. *Proc. Instn Mech. Engrs*, 194, 301 and 308.

Sherwin, D.J. (1983). Failure and maintenance data analysis at a petrochemical plant. *Reliab. Engng*, 5, 197.

Simpson, H.G. (1971). Design for loss prevention – plant layout. *Major loss prevention in the process industries* (London: Instn Chem. Engrs), p.105.

Smith, D.J. (1985). *Reliability and maintainability in perspective*, 2nd ed. (London: Macmillan).

Sofyános, T. (1981). *Causes and consequences of fires and explosions on offshore platforms. Statistical survey of Gulf of Mexico data.* Det Norske Veritas, Rep.81–0057.

SRI Int. (*1988a*). *Directory of Chemical Producers – Canada.*

SRI Int. (*1988b*). *Directory of Chemical Producers – United States.*

SRI Int. (*1988c*). *Directory of Chemical Producers – Western Europe.*

Summers-Smith, D. (1982). Leakage of flammable vapour through dynamic seals. In IEE (1982b), *op.cit.*, 7/1.

Sutton, O.G. (1934) Wind structure and evaporation in a turbulent atmosphere. *Proc. Roy. Soc.*, A146, 701.

Sutton, O.G. (1950). The dispersion of hot gases in the atmosphere. *J. Met.*, 7, 307.

Veritas Offshore Technology (1988). *Worldwide offshore accident databank* (Hovik, Norway).

Wagner, W. (1973). New vapour pressure measurements for argon and nitrogen and a new method for establishing rational vapour pressure equations. *Cryogenics*, 13, 470.

Wallace, M.J. (1979). Controlling fugitive emissions. *Chem. Engng*, 86 Aug.27, 79.

Welker, J.R. and Schorr, H.P. (1979). LNG plant experience data base. *Am. Gas Ass., Transmission Conf, paper 79-T-21.*

Wetherold, R.G., Rosebrook, D.D. and Tichenor, B.A. (1981). *Assessment of atmospheric emissions from petroleum refining.* Environ. Prot. Agency, Ind. Environ. Res. Lab., Rep. EPA-600/S2–80–075 (Research Triangle Park, N.C.).

Wetherold, R.G., Harris, G.E., Steinmetz, J.I. and Kamas, J.W. (1983). Economics of controlling fugitive emissions. *Chem. Engng Prog.*, 79(11), 43.

Wheatley, C.J. (1987). *Discharge of liquid ammonia to moist atmospheres – survey of experimental data and model for estimating initial conditions for dispersion calculations.* UK Atomic Energy Authority, Systems Reliability Directorate, (Culcheth, Warrington, Lancashire), Rep. SRD/HSE/R 410.

Wiekema, B.J. (1984). Vapour cloud explosions – an analysis based on accidents. Pt 1. *J. Hazardous Materials*, 8, 295.